OIL ON THEIR SHOES

PETROLEUM GEOLOGY TO 1918

BY
ELLEN SUE BLAKEY

PUBLISHED BY
THE AMERICAN ASSOCIATION OF PETROLEUM GEOLOGISTS
TULSA, OKLAHOMA, U.S.A.
1985

(Frontispiece)
The first geological field party in
Oklahoma Territory struck camp in the
Gypsum Hills of Blaine County in August
1900. It consisted of Paul J. White (left),
Roy Hadsell and Charles N. Gould (right).
Western History Collections, University
of Oklahoma.

Library of Congress Cataloging-in-Publication Data

Blakey, Ellen Sue.
Oil on their shoes.

    Bibliography: p.
    1. Petroleum--Geology--History. I. Title.
TN871.B488 1985   553.2'82'09   85-20021
ISBN 0-89181-803-0

Editors: Fred A. Dix, Robert H. Dott, Sr.,
         Ronald L. Hart, Anne H. Thomas
Production Manager: Ken Frakes
Typographer: Eula Matheny

Contents

· DEDICATION ·

When we began this work in 1983, we contacted as many of the old-guard petroleum geologists as we could locate, asking for photographs and stories. Some of them had long since retired and we had no records of them. It took a little digging, but once we found them, they overwhelmed us with their eager response.

When we explained that we only wanted materials prior to 1940, a number of people were upset: they felt that later material might well be lost—just as much of the earlier material had been—if something were not done about it soon.

They were right. We were thinking too small to begin with—and the material we received showed us that. For that reason, we decided to do a three-volume set instead of the one original volume planned.

One of those who responded was Ottmar F. Kotick who wrote us, "Many of the old guard are dead. Ergo it's high time for (a) history before the rest of us hunt new oilfields in wherever we are headed."

Kotick's remarks were appropriate. While we have been preparing the materials, at least one of the people who provided us with material went to hunt those new oilfields. We would like to dedicate this volume to those who have helped by providing their stories and photographs to be preserved for future generations—and to those who never had an opportunity to tell theirs. They have our gratitude—and may there always be an interesting horizon wherever they go, and an honest boulder for a pillow.

Most of the stories told here were supplied by geologists and their families. Some were from their letters, diaries and geological accounts. Others came from tapes and reminiscences. Still others came from works which they or others had published. Early newspaper accounts and company documents added still another dimension. But like all reminiscences, they are colored by personal points of view—by individual preferences and prejudices, by which side of the mountain or valley they happened to be standing on at the time. One geologist assures us that only his version is correct; another with an entirely different point of view vehemently responds that the first story could not be further from the truth. Those of us who only sift through words and listen cannot judge. We can only report. And if there are other versions we have not told, we have remained silent without malice but because we have not found those who told a different tale. Some very fine geologists left no personal accounts. Others' accounts are still buried with families we were never able to reach. So the stories we tell here may not be the most important geologically. Even though some may differ from what many believe to be factual, they certainly are indicative of the experiences of the early-day geologists in their quest for oil and gas. Our gratitude goes to those who responded to our request for materials. And our apologies go to the memory of others, those whose rich tales we never heard or found.

Young would-be geologists take samples along a hillside, at the turn of the century. George Hansen Collection, Brigham Young University.

OIL ON THEIR SHOES

OIL ON THEIR SHOES

Back around the mid-teens, Jack H. Lowe of the Sidney, Nebraska *Telegraph* described the geologists as he saw them—men who had often frustrated him in his efforts to get a good story.

"If you see a man walking down the street with oil on his shoes, where it shouldn't be, and no oil on his hair, where it should be, that's an oil man. If he has a faraway look in his eye and seems to be contemplating the depth of the first Jurassic sandstone in Persia, that's a geologist.

"Have pity on him. He's just as lonesome as he looks. He'd love to tell you everything he knows, but he doesn't know how. When he greets St. Peter at the gate and is asked to give an accurate account of his life on earth, he'll start out by saying, 'Well, whatever I say I don't want to be quoted because you never can tell what might happen, but...' That, my friends, is a Geologist."

Petroleum geologists are a special breed. Walter Youngquist once wrote that "little boys who pick up rocks either go to prison or become geologists." They are "Boy Scouts who hated to give up camping when they went to college so they majored in geology." After all, he explained, what other business could an individual be in and go to the Grand Canyon or the Swiss Alps and claim he was working?

"A man simply cannot become a geologist," wrote newspaperman Lowe, in the Midwest's early days of petroleum geology. "He is born one, raised as one and educated as one. It requires a certain kind of temperament, a deliberate sort of speech and patience beyond measure. A geologist listens to more silly questions than any other human, and he must weigh each question and devise each answer with the greatest of skill. He must be able to talk at length without disclosing what is foremost in his mind. He must know how to be polite but firm, decisive but evasive, positive but negative, effusive but quiet....

"Furthermore, a geologist must know what is going on down in the depths of Mother Earth, and that is something we always thought was reserved for the Lord."

Some geologists, however, view their work in more philosophical terms. In *Beyond the Observatory*, Harlow Shapley wrote, "There are within us the same chemical elements that make up the mountains, the pine trees, the seashore....We are indeed of the earth—brothers of the boulders, cousins of the clouds and distant kin, by way of chemical tie-up, of the fossil plants and animals." All these entities form "our animate and inanimate brotherhood."

This kinship and the geologist's understanding of the relationships of the earth have helped him as he moved along the edges of accepted knowledge. John G.C.M. Fuller of Calgary, Canada, called this understanding a "geological attitude"—a "particular point of view which has been a characteristic of geologists since they first tracked mud and heresy into the schools of learning some 200 years ago."

And if they were often ostracized or misunderstood, it was frequently because the thoughts and concepts embodied in their discoveries often tested the ideas so long and dearly held by church and elder scientists concerning time, man and the balance of nature itself.

"Man, to survive and progress and to understand in manways, founded the science of probing beneath this earth in search of minerals to be put to use," explained NBC newsman Frank Blair in 1966. "From this desire to understand and to explain the probing by man, the science of geology came into being and as a fruit of this science the civilization of man has benefited and progressed rapidly."

Some feel that this progress was the result of great scientific curiosity and endeavor; others say it was the result of the spirit of adventure. Still others claim that it has been the result of outside pressures. According to Fuller, geological activity first takes place mainly in response to industrial and social pressures. Petroleum geologists have probed, discovered

(Preceding page)
Oil seep on the bank of the Colorado River at Bennett's Oil fields in Garfield County, Utah, in August, 1915. E.C. La Rue, U.S. Geological Survey.

and invented the tools with which to carry on the work as the need has arisen.

Whatever the reason, the outcome has been fortuitous for civilization. When the petroleum industry had its beginning less than 150 years ago, machinery accounted for one-third of all work performed in the United States. Men and animals performed the rest. By the 1960s, more than three-fourths of all the energy required by America was provided by petroleum products. The petroleum geologist was at least partially responsible for this change. "The geologist is the keystone of a great arch involving not only his own but many other industries," wrote Lewis George Weeks in 1976. "No other group has done more to raise living standards worldwide....Less than 10,000 active petroleum geologists, and lesser numbers in the past, have pointed the way to trillions of dollars worth of energy for the benefit of mankind."

Thus petroleum geologists—for all their study into the ancient history that formed and deposited underground riches—have been—and still are—in the forefront of science. They study the past in an effort to serve the future. They may battle bandits and insects, snakes and vultures, high mountain ranges and barren deserts. They may kiss their women goodbye—and sometimes bundle them along—in search of black "goo" or "funny little anticlines." They may share their last bit of food with industry rivals, but they are still human—and liable to hedge about the depth of their well or the lay of the land when the rivals get too close. For all their foibles—and for all the technological advances—they still toss the dice with Mother Earth and dream of sevens.

·ROCK OIL AND ANTICLINES·

From the dawn of history, man has used petroleum. Oil which seeped to the earth's surface was used to caulk boats or to cure all manner of ills. Women enhanced their eyes with an exotic eye shadow made from asphalt. Before 500 BC, Babylonians waterproofed their roofs with bitumen while Egyptians embalmed with it. Syrians mixed heavy oil with wood ash for cooking and heating. Greeks soaked projectiles in flaming oil ("Greek fire") in warfare and set seas aflame by igniting oil scums during naval battles.

On the North American continent, aboriginal native tribes drank oil to cure fevers. The "rock oil" was considered a magic potion to cure burns and skin diseases, sprains, rheumatism, coughs and stomach ailments. The Karankawa Indian tribes in what would become Brazoria County, east Texas, gathered to drink the blue-black water that collected in the rock depressions and to bathe in what they called 'sour dirt.' The black goo was used to pitch war canoes and woven baskets which could then hold water. Hides—and later blankets—were soaked in the oil springs, and the small amounts of petroleum retrieved were shared with family or tribe.

Early American colonists quickly adopted the medicinal cures of the natives. George

Washington surveyed the west in 1753 and purchased a burning spring in western Pennsylvania and an oil outcrop on the Kanawha River in present-day West Virginia. In 1783, troops used petroleum from Oil Creek, Pennsylvania, as treatment for rheumatism and as a purgative. Yankee peddlers sold it for everything from removing warts to easing childbirth. Trappers in the West and Southwest used it to rub on the galled backs of their horses—and of course what was good for the horse usually went double for the man.

Meanwhile, in the learned circles of Europe, interest in an infant science—geology—was growing. In 1785, James Hutton, a naturalist and farmer, spoke before the Royal Society of Edinburgh on the "system of the earth, its duration and stability." He talked of one constant in the universe—the inevitability of change and the inevitability of gradualness. In *Theory of the Earth*, he discussed the concept of Uniformitarianism—that "the present is the key to the past," that the history of the earth is dependent on a continuous cycle of weathering, erosion, sedimentation and uplift. This was a revolutionary idea and a challenge to the conventional religious thinking of the time. It did not at first draw much attention, however. A close friend of Hutton's, John Playfair, geologist and professor at the University of Edinburgh, commented, "It might have been expected, when a work of so much originality...was given to the world, interesting not to mineralogy alone, but to philosophy in general, that it would have produced a sudden and visible effect....Yet the truth is, that it drew their attention very slowly, so that several years elapsed before anyone showed himself publicly concerned about it, either as an enemy or a friend." Once it did start to draw criticism, Playfair published his own work in support of Hutton's ideas. Sir James Hall, another friend of Hutton, was the first to perform experimental studies in geology and was the first to melt basalt.

Although Hutton's work remained largely obscure except in the circle of his Edinburgh colleagues, it did inspire others. Geologists fanned out over the European continent confirming many geologic concepts and mapping structures. A second burst of scientific creativity in Scotland was inspired by such men as Sir Charles Lyell, Benjamin Peach, John Horne, Hugh Miller and Archibald Geinkie.

Peach and Horne were the first to describe and interpret an overthrust belt. Miller was a stonemason who later became a journalist and one of the better writers on geological concepts. Because of his deep religious beliefs, Miller was bothered by Darwin's theories. Finally unable to reconcile what he believed to be major conflicts, he committed suicide.

ES, VIRGINIA CITY, NEV: OCTOBER 28TH 1879.

Interest in all types of underground riches was widespread, from coal to oil to gold. General Ulysses S. Grant (center) poses in 1879 in front of a mine at Virginia City, Nevada.   Brigham Young University.

Oil exchanges sprang up throughout Pennsylvania as the fields mushroomed overnight. Many a man gambled his fortune on the prices of crude--and watched that fortune disappear when the shallow Pennsylvania wells tapered off and the oil exchanges crashed. The Parker Oil Exchange posed for the camera in 1874. McLaurin, Sketches.

Early geologic surveys of the West lived off the fat--or lean--of the land. The horses were turned loose to pasture on the native grasses. The deer head rests on a barrel by the table. The venison was most likely on the men's plates. The bones the dog got to gnaw on. National Archives.

·Boring For Salt·

In 1828, Thomas Ellison of Kentucky wrote Edmund Rogers of Mount Hope, Kentucky, confirming the purchase price of a Roan filley (sic). "We have no news except Col. Emmerson and Mr. Stockton has been boring for salt on 'Rennix,' " he added. "Struck a lake of oil. The well has been running for the last four days throwing out large quantities of rock oil. It would spout out at least 15 feet above the top of the ground as large a stream as a man's body perfectly pure. Cumberland River has been covered from bank to bank for three days. The river was this evening set on fire. It burnt for at least a mile in a stream or sheet of fire. I have no doubts but in the time it has been running, to speak in the bounds of reason, that five hundred thousand gallons of oil has (sic) run down Cumberland River and is still running, but not quite so fast. Col. Emmerson has barrelled twenty barrels of it. It burns well in a lamp and is good to paint leather and I have no doubt it will be a good medicine for many complaints, particularly the reumatic (sic) pains. The whole atmosphere is perfumed with it. It is a complete phenominon (sic). I could write a whole sheet and not say half, it is late and I am sleepy. I am yours, Etc."

·System Of The Earth·

It was still some distance away, but the Industrial Revolution was rearing its head. Newly constructed factories demanded coal and minerals; mining and quarrying operations accelerated. Colliers and quarrymen had understood the concept of stratigraphy for years. Now, naturalists and geologists promised help locating and identifying deposits. They were more concerned about the results—making their factories work—than the ideas which had governed accepted thinking in the past.

So much attention was stirred up with the new public awareness of geology that scientific speakers drew thousands. Adam Sedgwick, founder of the Cambrian System, was received enthusiastically by 4,000 people at Tynemouth, England—and for six weeks, 1,200 people turned out to hear him daily. Charles Lyell lectured in Boston in 1841 and had to give his lectures in shifts to keep the audience down to 1,500. When Sir Roderick Murchison, author of the Silurian System, was appointed director of the Geological Survey of Great Britain, members of the House of Commons rose to their feet and cheered.

While "pure" scientists were debating theories of geology, the more practical Americans were busy creating new markets for petroleum products. An enterprising doctor by the name of White bottled the oil from the Kentucky well and sold it for a cure-all. He even developed a market for it in England, and as a result became wealthy. Samuel M. Kier also ran into oil while digging salt wells in the 1840s and decided to take advantage of the natural troublemaker. He set up an operation to distill petroleum at Tarentum, Pennsylvania. He put the crude oil into a five-gallon drum and lit a fire underneath to

(Preceding page)
When oil was discovered in Pennsylvania in the 1860s, thousands of men and women flocked to the state. Many fields were floor-to-floor derricks, as was Benninghoff Run, Venango County. An hour after the photograph was taken, the derricks were destroyed by fire started by lightning. McLaurin, Sketches.

evaporate the volatile gasoline. Then he bottled the remainder as an elixir and sold it for 50 cents a half-pint. Kier's Rock Oil was touted as an all-purpose medicine to cure burns, ulcers, cholera, asthma, indigestion, rheumatism and blindness. At first, Kier sent salesmen in gaudy red wagons throughout the Eastern towns and hamlets. An elaborate circular with drawings of the noble red man touted his product:

> The healthful balm, from Nature's secret spring,
> The bloom of health and life to man will bring:
> As from her depths the magic fluid flows,
> To calm our sufferings and assuage our woes.

But Kier misjudged the market. His little red wagons were not cost-effective, and after two years, he gave them up and only sold wholesale to druggists.

·A NEW SCIENCE·

Sir William Logan, director of the Geological Survey of Canada, first became interested in geology in 1831 when his uncle in London sent him to the mining district of Wales to learn about the copper-smelting process. He wrote his brother, Henry, to pack his trunk with old clothes and "some good work on mineralogy and geology and if it will hold any more, put in some of my flute music." By 1843-44, Logan had graduated to head of the Canadian Geological Survey. One theory he developed settled a question which had troubled geologists for years: coal, he insisted, was formed where it was found—not by accumulations of drifted materials. Logan went to the Gaspe area to find coal deposits in 1846. But he also found two petroleum springs near Gaspe Bay on the shores of the peninsula near the mouth of the St. Lawrence River. There the rocks were thrown into a series of gentle undulations he called anticlines. Most of the oil springs occurred along the axes of the anticlines. Logan's observations were probably the first records of the anticlinal or structural accumulation of oil.

In Massachusetts, Edward Hitchcock was making important strides with his geological surveys of the state. A methodical man, Hitchcock published an explanation of the new science—*Elementary Science*—which was widely read.

While some men were interested in the theories and the reasons behind mineral resources, others were finding new uses for them. In Scotland, James H. Young extracted "coal oil" from shales near his home; and in Vienna, Adolph Schreiner invented the modern kerosene lamp. Dr. Abraham Gesner, a Canadian chemist and former provincial geologist, received a U.S. patent to manufacture illuminating gas. By 1850, 56 plants supplied coal-gas to urban areas. Justus Liebig, professor of chemistry at Germany's University of Giessen, predicted that the oil needs of the future would come chiefly from mineral sources rather than from whales or animals.

On the West Coast, most citizens rated oil as second class compared to gold. Geologists and surveyors felt differently, however. In the 1850s, Lieutenant R.S. Williamson led a survey party in the San Joaquin Valley and concluded that California was ready to enter the oil and asphalt industry. "Bitumen is par excellence the mineral of Southern California, being found in almost every county south of San Francisco," wrote Thomas Antisell, geologist for Lieutenant J.G. Parker's survey from San Diego to San Jose in 1855. The state geologist advocated using natural bitumen to manufacture illuminating gas. By 1856, the San Fernando mission was lit by natural gas, and the Stockton courthouse well supplied natural gas for the town.

By 1859, the lamp trade alone burned seven million gallons of kerosene annually. Prices for crude oil rose to 7.5 cents a gallon. Instead of the old quart cans or gallon jugs which Americans had used to transport crude to refineries, they began to fill whiskey kegs—and thus the 42-gallon whiskey barrel became the standard sales unit for crude oil. As prices increased and uses and markets grew, interest in the dynamic new science known as geology made its way through scientific, business and educational communities.

Many men took to the field with only the vaguest notions of a new science brewing in their heads. These amateur geologists still knew so little of petrology that they could not tell one kind of rock from another. Nearly every operator was ready to discourse learnedly on rocks, formations, strata, shales, sandstones or anything else that might come along.

·LINING OUT THE LAND·

The need for knowledge about the entire continent and its mineral resources was growing. The vast American West was just opening, and no one really knew what was out there in the area they dubbed the Great American Desert. Railroads were inching their way across the continent—and they needed fuel to move. Although coal was plentiful, some businessmen wanted to know more about other mineral deposits—such as gas and oil—which now appeared to have possibilities.

In the 1850s, Ferdinand Vandeveer Hayden, a young medical graduate, began exploring the West with little more than a pick and a shoulder bag to collect geological samples. Indians in the trans-Missouri area where he ranged thought he was crazy. Once, they surrounded him and dumped the contents of his bag onto the ground. All they found inside were rocks. The named him "man-who-picks-up-stones-running" and let him go. Hayden continued to dart from one dry run to another, climbing buttes and escarpments for no apparent purpose.

But what appeared madness to the Indians was a dream to Hayden. He envisioned a great geological survey which might "lay before the public such full, accurate and reliable information...as will bring from the older states the capital, skill, and enterprise necessary to develop the great natural resources of the country."

While Hayden was investigating the West, others were taking a second look at the East and Midwest. In 1858, Henry D. Rogers, geologist of Pennsylvania, claimed that the Pennsylvania oil fields were actually on anticlines rather than in crevices as formerly supposed. In 1861, Professor E.B. Andrews of Ohio traced a line of uplift from East Washington County, Ohio, to a point beyond the wells of the Little Kanawha River. It clearly showed how oil fields, and oil and gas springs, were located along the line. His maps and cross-section of the oil field may have been the first ever published. In 1865, Professor G.C. Swallow, Kansas State Geologist, and Professor B.F. Mudge authored the *First Annual Report on the Geology of Kansas*. It was noteworthy for its concepts concerning petroleum. Swallow limited his remarks to the eastern counties, claiming that the Indian uprisings in western Kansas put a damper on geologic undertakings in that area. Mudge claimed that oil floating in springs did not indicate the quality of the oil beneath the ground. The best material, he averred, would rise, leaving behind only heavier and impure elements. The report did not deter the more imaginative of the oil-hungry men, however.

After the Civil War, interest in the West and its minerals grew stronger. Several wells were drilled at oil springs in Nacogdoches County, Texas, near present Chireno, but production was small. But glowing reports depicted a veritable Garden of Eden further West. A "Captain" Sam Adams wrote the Washington D.C. *Chronicle* claiming to have explored the Colorado River (freely navigable for 600 miles from its mouth). One valley which he dubbed "Paradise Valley" bore wild oats, timothy, clover and seven-foot-high rye. Veins of gold, silver, copper and lead could be seen in the canyon walls or far up the side canyons. "We saw oil floating on a number of small streams entering the main river," he boasted. "Occasionally a spot could be seen on the land, composed of a sticky, tarry substance, where we found that a number of birds and squirrels had perished in endeavoring to extricate themselves."

But investors needed more than just glowing stories, promises and a fistful of ore specimens. They wanted reliable maps and careful appraisals. Many states began to hire geologists. The "crazy" Hayden was one of them.

Hayden became geologist in charge of the Nebraska Geological Survey, funded by a small sum of money left over after expenses of the territorial legislature. It sounded like play to Hayden. He was charged with examining "all the beds, veins and other deposits of ores, coals, clays, marls, peat and such other mineral substances as well as the fossil remains of the various formations."

Hayden lobbied Congress the next year for more appropriations; and in 1869, he found himself in charge of the U.S. Geological Survey of the Territories, covering Colorado and surrounding areas. His reports were always filled with superlatives and couched in terms of economic advantage. When he thanked members of the Union Pacific, Kansas Pacific and

Wildcatters hit it big in Indian Territory
with the Nellie V. Johnstone well.
Phillips Petroleum.

Denver Pacific railroads for their support, he reminded them that they were the real beneficiaries of his efforts. "The generosity on the part of such corporations towards men who are devoted to the advancement of knowledge or the good of the world, may be regarded as the index of their tone and character," he wrote.

One of the major changes during this period was improved topographical methods, first employed by the California survey in the 1860s. Until that time, the general method for mapping the country was to run traverse lines and "sketch everything in sight," according to Henry Gannett. "In its topographic work (the California) Survey made the first wide departure from (these methods). They violated precedent in three important respects—they surveyed areas instead of lines; they controlled the work by triangulation instead of traverse lines, and they sketched the topography from elevated stations. They sketched in notebooks. The triangulation was computed and platted in the office, and the sketches were transferred to the locations. These methods were carried to the 40th Parallel Survey, and thence to the Hayden Survey."

When Gannett joined the Hayden Survey in 1872, the primary triangulation work was done "with a ten-second theodolite, and the secondary work with a minute-reading instrument. Heights were measured by barometer and vertical angles, or a combination of the two."

Interest in the survey work increased in the 1870s as the result of the efforts of Hayden and others, such as Clarence King. King was a flamboyant young geologist who had graduated from Yale in 1862. He spent three years with the Josiah Whitney Survey mapping California and was appointed director of the U.S. Geological Survey of the Fortieth Parallel in 1867. The survey was to cover a 100-mile-wide cross-section of the continent from the Sierra Nevada to the Rockies roughly following the construction of the Central Pacific Railroad.

Until that time, geology had been mainly a field science. But King hired young laboratory geologists whose application of chemistry and physics to the study of rocks revolutionized the discipline. He was the first American geologist to bring the physical sciences of chemistry and physics to bear on geological questions.

His work, *Systematic Geology*, detailed his comprehensive theory on the evolution of the basin-and-range system along the fortieth parallel. The next year, King was appointed director of the newly created U.S. Geological Survey and charged with systematically classifying federal lands west of the Mississippi.

Meanwhile, in 1872, Ferdinand Hayden was hired to compile a massive topographical and geological investigation of the Yellowstone-Teton area. The report—complete with the magnificent photographs of William H. Jackson—did much to increase interest in the West, establish the national park system and create ever greater interest in geology.

The Powell Survey in Utah used the same methods for topographical mapping, according to Henry Gannett, but they added what he called the most outstanding improvement in topographical work of the time—the (Johnson) planetable which was "not only (used) as a sketching board but as an instrument for triangulation."

In 1882, Henry Gannett became chief geographer for the U.S. Geological Survey. "The early topographic work of the Geological Survey was, frankly, of a reconnaissance character, and it was intended to be," Gannett explained. "We were trying to get the work of making a map of the country started; we were trying to induce Congress to make continuous appropriations for it....Reconnaissance work was deliberately decided upon with its low cost and rapid progress, until such time as Congress could be counted on for more liberal support for better work."

·DIVINING RODS AND DOODLEBUGS·

While geologists and other scientists analyzed the land in terms of new theories, hucksters and hustlers moved rapidly to reap the rewards of the new riches that appeared possible with petroleum.

Pennsylvania was the site of the early Eastern fields. In 1859, George H. Bissell and Jonathan Eleveth, two New York lawyers, and Professor Benjamin Silliman, Jr., organized an oil company and purchased 105 acres at Titusville and hired an unemployed railroad conductor named Edwin L. Drake to drill a well. When Drake finally succeeded in spite of himself, the rush was on.

The salt industry had developed technology which was adaptable for oil, and the boom blossomed quickly. Towns sprouted up in abundance with all the wickedness that boomtowns seemed to attract. One of the most notorious was Pithole, which, it was declared, "emerged from the indignity of civic seclusion to the glory of a near-metropolis; with unprecedented rapidity, the uninteresting Pennsylvania barrens became a hub of nearly twenty thousand victims of oil-neuroticism. Homes, endowed with all the conveniences which modernity then embraced, sprang up like mushrooms; hotels and taverns, architecturally lavish in keeping with the tastes of people who could well afford lavishness, were born at the rate of one every day and a half....The town became the mecca for the speculator, the adventurer, the shyster. The municipality absorbed a daily deluge of post-war unemployed, who came to Pithole to start anew and encountered no difficulty....Hotel lobbies hummed with talk of oil. Each day brought in reports of astounding productivity. Oil was struck everywhere, and there seemed no limit. One operator, better favored with foresight than scrupulousness, disposed of seventeen-sixteenths of his property."

Shaffer grew from a community of one house and one barn to 3,000 energetic souls, in sixty days. Boarding houses were literally erected overnight—the boards were assembled

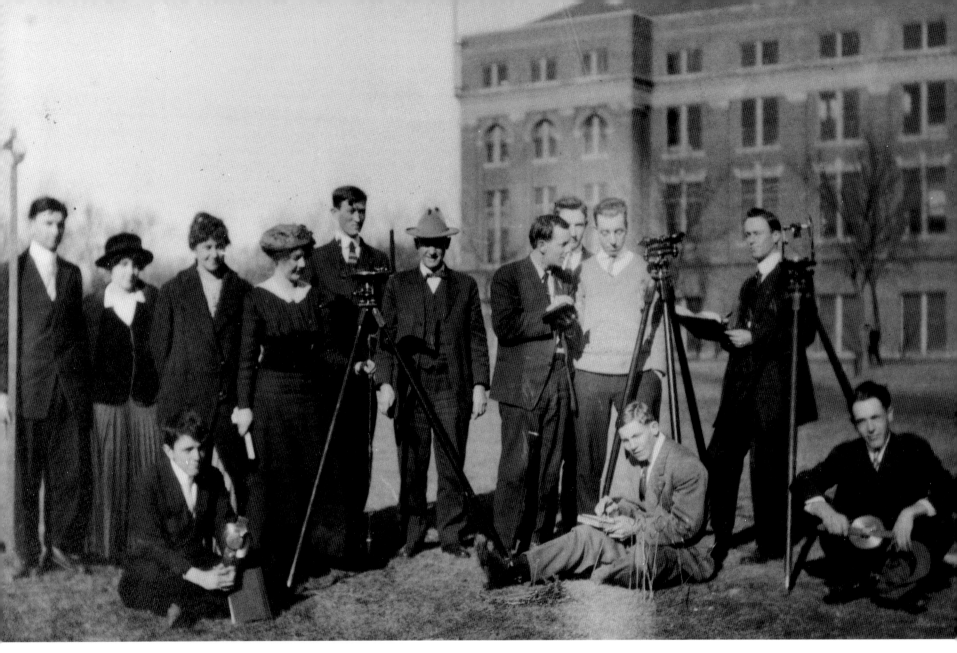

Kansas students pose for the camera around the turn of the century. Blakey Group Collection.

without a frame, which speeded construction but resulted in considerably less comfort or protection from the elements.

At first, picking a drilling site seemed relatively easy: Simply choose a spot where oil signs were prevalent—oil seeps or springs. It did not take a geologist to do that. But it did not always work. Drilling at the oil seeps at Seneca Oil Spring in New York in 1857 had yielded nothing. Some better way had to be found; and men with fast hands and smooth voices moved in quickly to take advantage of that weakness.

Some of them swore by "creekology." Since oil often coated the surface of streams or springs, creekologists claimed petroleum could be found beneath the riverbeds or in the curve of rivers or creeks. Others claimed it would be found near cemeteries (usually on high ground), while others insisted drilling should never be near a sawmill (characteristically in low land). The real geological reasons were often unknown, but the quick turn of phrase—"near cemeteries but not sawmills"—caught on.

Some trusted to dreams. Along Oil Creek in Pennsylvania, a man by the name of Kepler dreamed he was shot at by an Indian with a bow and arrow. A coquettish woman handed him a gun to frighten away the redskin. At that point, oil gushed from the ground. When he visited his brother on the Hyde and Egbert farm, he identified the place he had seen in the dream. He secured three partners, leased one acre at three-quarters royalty and drilled the Coquette well. It came in at 1,200 barrels a day and produced 800 barrels daily for a year.

Some turned to mediums. A spiritualist by the name of James was riding calmly in his buggy in the neighborhood of Pleasantville, Pennsylvania, when he was suddenly "agitated by a spirit-guide." He seemed to be forced out of the buggy, over a fence toward the south and then the north end of a field. He was thrown to the ground where he made a mark with his finger and forced a penny into the earth. James drilled at the site, and the Harmonical well came in at over 100 barrels a day. For a while, James was known as the "boss locator," but the well slowed down; and eventually, he left a trail of dry holes through the state.

Jonathan Watson, also of Pennsylvania, relied on his own personal medium—his wife—and occasionally let her locate one of his prospecting holes. Eventually, he began using other mediums. But as the spirit-magic wore off, he turned to another ancient technique—the divining rod. The doodlebug or water-witch stick had in past times been used to find coal, iron, gold, buried treasure, water and whatever the current rage demanded. Watson hired two "doodlebugs"—men with the divining gift—but their first locations proved dry. When they claimed the field was in the middle of a river, Watson ordered the stream dammed and diverted. The process took several months. When the wells were drilled, they all came in dry.

Many claimed to follow "close geology." If a man was close enough to smell oil from another well, he was in a good spot to drill. Some developed elaborate gimmicks such as the black box—also dubbed a doodlebug—which showed up in a number of early fields even as

The 1873 U.S. Geological Survey party, shown here, included Frederick Hayden and helpers Stevenson, Holman, Jones, Gardner, Whitney and Holmes. The party made the first ascent of the Holy Cross Mountain. William H. Jackson, U.S. Geological Survey.

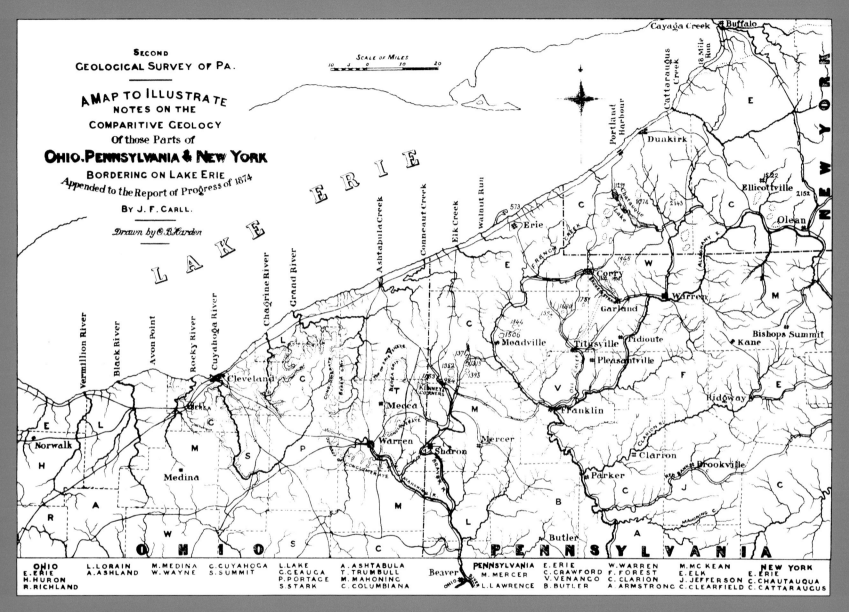

The Geological Survey of Pennsylvania was the largest and most active state survey in the United States during the 1870s. It probably originated the concept, though not the name, of stratigraphic trap, for that was the mode of trapping the oil in the Verrango Sands of the Pennsylvania oil fields. The 1875 report recognized that fields in Ohio and West Virginia occurred on anticlines. Courtesy, Jeffrey Heyer.

As the photographic arts developed, early explorers often took photographers, such as these at Ausable Canyon, New York, to record the wonders of the land. George Hansen Collection, Brigham Young University.

late as the 1930s. One memorable machine was outfitted with electrical wires, dials and bells. It was transported across the potential field in a shrouded sedan chair, carried by four men. When the bells started clanging, the bearers stopped and marked the spot for drilling.

Many employed the jargon of the developing science of geology, no matter how little they knew. "The Petrolians (Pennsylvania) are nothing if not geological," wrote one observer. "Nearly every operator is ready to discourse learnedly on rocks, formations, strata, shales, sandstones (comprising every thing from limestone to conglomerate). As in nature, so in human nature—no two agree." If one claimed the best wells were to be found on the east side of all runs and creeks, another would insist that prudent men only bored on slopes. Still another would claim that the rents in the hill-tops indicated the country was of

volcanic origin, and the petroleum was simply the smoke and cinders of the forges of the old Norse god Vulcan. It was a wise man who understood the dangers and the trickery and managed to keep a tight fist on his pocketbook.

One strike on the Hoover farm near Franklin was especially noteworthy. On a placid afternoon, drillers working on a hillside felt the tools drop at only eight feet. Immediate conjectures arose—there must be a tremendous oil pool beneath the hill for such an occurrence. The workers drew up the tools and found them dripping—not with oil but with an amber fluid that smelled and tasted like beer. All pandemonium broke loose. Beer flowing like oil! A swarm of conveniently thirsty citizens rushed to the site while everyone dreamed of the number of breweries that would be put out of business by this new development. Samples were being passed around to appreciative bystanders when a rotund German rushed up. They had not discovered beer flowing under the earth as they assumed. They had tapped into the storage tun of a local brewer. He had cut a tunnel into the hillside and installed his beer inside to keep it naturally refrigerated. Workmen who had gone to the tun to obtain a load of brew had been startled to see the pipe dangling in the vat and the fluid fast disappearing. The drillers—their dream of riches suddenly shattered—withdrew the pipe while the disappointed crowd stomped off. The drilling was moved to another hillside, but it was almost an anticlimax. After all, it was only oil.

·No More Candles·

Canadians were as oil-hungry as Americans. In 1858, the *Sarnia Observer* reported the discovery of an "abundant supply of mineral oil" in the township of Enniskillen. The writer noted that when the oil burned, it emitted, "on account of the impurities, a dense black smoke;" but learned sources were sure that upon purification it would make "a splendid lamp oil." A month before Drake completed his well, James Miller Williams had set up refining facilities at Oil Springs and established his oil on the market. It could be purchased from an agent in Sarnia for a hefty $1.25 a gallon. "No one who gives it a trial would ever think of returning to the use of candles," they reported. By 1860, the company had become the Canadian Oil Company. Two years later, they received two gold medals at the International Exhibition in London—one for being the first company to commercially produce crude oil and another for the first to refine oils in Canada.

By mid-1861, Enniskillen had grown to 1,600 people. There were 100 producing wells with 300 more being drilled along the banks of Black Creek. There were several small stores and the usual overcrowded hotel. The village's initial water well was contaminated by oil. The small, sluggish-flowing stream behind the hotel was used for everything. Diggers, covered with oil, washed themselves after the day's work, swilled the mud off their boots and quenched the thirst of their horses. It also provided water for tea, coffee and

Early day geologists gather at a special occasion that has since passed into obscure history. Seated left to right: L.M. Prindle, A.H. Brooks, J.E. Spurr, J.A. Taff, H.B. Goodrich. Standing, left to right: T.W. Stanton, G.O. Smith, C.W. Purington, G.W. Stose, A.C. Spencer, (?) Lord, G.H. Girty, F.C. Schrader, T.W. Vaughan, (?) Dorsey and G. W. Tower. U.S. Geological Survey.

boiling the fat salt pork, the staple food in winter months. Beds were in such shortage that hotel guests who retired early were awakened at midnight "so that other gentlemen might take a sleep." Undressing was not considered strictly requisite, but one American overdid things when he went to bed in his boots. Despite the conditions, there was little lawlessness—no knifings, shootings or fights. Contractors charged $2 to $4 a foot to sink a well, depending on the depth, and it was said that the average cost of a completed oil well was $300.

In 1861, a 50-year-old Irishman by the name of Hugh Nixon Shaw decided to abandon his general store at Cooksville and try his hand at oil. He had only $50 working capital and worked alone. His capital was soon exhausted, however, and so was his credit within a 20-mile radius of Oil Springs. Still he stuck to it. He dug a well 4 × 5 feet, through 50 feet of clay, then bored through 158 feet of rock with a spring pole. On January 16, 1862, oil rushed up from the well and overflowed at the surface. It was the first gusher Oil Springs had seen. Shaw tried to bring it under control with a bag filled with flax seed, but the pressure forced the bag back to the surface. A second bag stayed down a little longer but also came up. Oil spewed 20 feet into the air for three days and nights. Thousands of barrels of oil spilled across the surface of the land and covered the ice of the creek three and four inches deep.

Before the year was up, 30 gushers had been drilled within one square mile. Shaw did not live long enough to enjoy his windfall. On February 11, 1863, he lowered himself into an oil well to pull up a piece of pipe. He was within 15 feet of the surface when he was overcome by the gas fumes. He fell back into the oil and disappeared.

·HUNTING BETTER ODDS·

One Pennyslvania pioneer who used logic and reason to find oil in the 1870s was Cyrus D. Angell. Angell had made money in oil through sheer luck, but he wanted better odds. He felt that oil existed in continuous sand bodies rather than just crevices as so many were touting. If the data in two oil fields within the same region corresponded, it proved a continuous sand belt flowed between the fields. Drilling between the two areas, then, could uncover other fields in the same sand. To do this, individuals would have to compare the depths of the sands, the distance between them, thickness, texture, quality, quantity, color and gravity of oil in each field. He spent eighteen months and $60,000 for land and leases before he was able to start his first well to test his theory. This trend-play method of oil prospecting paid off. More important, it demonstrated that finding oil was more than just luck.

There was some activity in the Mid-Continent area. A small field was opened near Nacogdoches, Texas, in 1877. Up in Indian country, the Chickasaw agent had noted that the Indian oil springs were attracting considerable attention. "They are said to be a remedy for chronic diseases," he wrote. "Rheumatism stands no chance at all, and the worst case of dropsy yields to its effects. A great many Texans visit these springs, and some people from Arkansas. They are situated at the foot of the Wichita Mountains on the Washita River."

Despite the increased interest in oil, petroleum geology was still looked on with some dubiousness and raised eyebrows. It might be 'smart' to know something about scientific theories, but it was still not respectable (or lucrative) for a young man to become a scientist—a doctor, perhaps, but certainly not a geologist—at least not in America.

·Dance By The Light Of The Well·

In the 1880s, most people thought "oil country" was Pennsylvania, New York, Ohio and West Virginia. Although California was producing enough oil to be mentioned in government statistics, it was far away, and no one really suspected the splendor of its petroleum riches. In between was Kansas.

Prospectors, it was said, could easily be spotted in Chanute. They were "those neat, clean, well-dressed young-old men recognizable all over the American business world as those who have done things. They are...above average size, are gray-headed or at least grizzled, and have waists that bar them alike from the waltz and ready-made trousers. These are the kind of men who are finding the places, making the leases and boring the wells of the present Kansas oil field." They were the men, like G.P. Grimsley of Topeka, who claimed that the best, if not the most scientific way, to ascertain the presence of gas and oil in central Kansas was to drill.

In 1882, the Kansas Oil, Gas & Mining Co. began drilling east of Somerset in Miami County. They were hoping to strike oil. Someone decided to help out a little, and during the night poured several quarts of crude oil down the well bore. Naturally there was an excellent "oil showing" the next day. The news spread fast, but one of the company members, John J. Werner, was suspicious. His suspicions were confirmed when another company member confided in him. He pointed out that it would be easy to sell the well to more gullible individuals just as it stood—and at a handsome price. Werner conceded that it would bring a handsome profit, but he refused to have anything to do with the swindle. The man who had salted the well withdrew from the company shortly afterward. (It was said that he became wealthy through operations in which he was able to give his talents freer play.) Werner went on drilling. In July 1882, they struck gas. Residents of the area were thrilled with the discovery. But there seemed no way to immediately turn the discovery to commercial profitability—at least not until two members of the company, H.J. Foote and Charles Kitchen, both of Olathe, hit upon a plan. The two built a dance platform in an elm grove near the well. They then ran a line from a 20-foot vertical pipe at the well about eight feet above the platform and installed flares. At one edge of the platform they built a refreshment stand. Two young musicians from Paola—about seven miles away—were hired, one to play the violin, the other the violincello. Everything was ready for what became known as the Westfall gas well dances.

The platform was big enough to accommodate six sets in the old-fashioned square dances. In addition to the regular square-dance fare, there were occasional schottissches, polkas and waltzes. Young couples—usually in Sunday best—drove to the well in buggies, buckboards and other horse-drawn conveyances. The enterprise was an instant success. Some nights, 500 people showed up to shuffle across the dance floor. As many as four attendants were busy serving lemonade, candy, sandwiches and cigars.

Down south of Kansas, the Choctaw and Cherokee nations had begun to think about the riches that might lie beneath their land. They formed their own oil companies, and Dr. H.W. Faucett of New York was hired to drill. They began 20 miles west of Tahlequah; but the well turned up dry. Another group prospecting along a branch of the Clear Boggy River 14 miles west of Atoka had no better luck. When they drilled in 1885, they got nothing but dust for their trouble.

The news of the Kansas strikes reached all the way to Pennsylvania and New York where oil men were having a tough time. On May 16, 1884—Black Friday, as it was known then—the Bradford, Pennsylvania oil exchange had crashed, leaving many men who had gambled on the rise and fall of crude prices with little but a handful of change. One of them was William M. Mills. When Mills read of the gas strikes in Kansas, he and his wife struck out for Kansas. He spent two years, living on borrowed money, driving from one spot to another searching for convincing indications of oil. When he read about a gas well being brought in near Paola, he hurried to the town and began signing up leases. Although times were hard and Mills had dozens of setbacks, he managed to set up drilling. Mills thought, however, that he was drilling for gas. His No. 1 Norman finally paid off in 1892.

"November 25," Mills wrote in his diary. "Drilled through oil sand. Got more oil, but no more gas. I worked at well all day and night.

"November 26. Worked at well in morning. Bryson and I took 10 o'clock train for Osawatomie. Shipped 520 feet of 4 1/2-inch casing to Neodesha.

"November 28. Well commenced flowing oil at noon. I stayed with it all day and night. Gathered 60 gallons of oil and shipped it to Osawatomie."

The No. 1 Norman was the first oil well west of the Mississippi and east of California to yield oil in commercial quantities. Its significance was more apparent later, and across the page with the November 28 entry, he scrawled in big characters with a red pencil, "Oil! This was opening up of Kansas oil field."

The well was capped and the drilling equipment removed although the derrick was left standing. On the afternoon of January 29, 1893, a windstorm blew it down, snapping off the drilling nipple which projected above the floor. Oil began to flow out and trickle into

Photographers on the U.S. Geological Survey pose in an 1870 stereoscopic view. As with the other "necessities" of their life, dark-room equipment had to travel by pack horse.   William H. Jackson, U.S. Geological Survey.

William H. Jackson, Dr. Peale, Dr. Turnbull
and Dixon pose in camp during their 1871
expedition for the U.S. Geological Survey.
William H. Jackson, U.S. Geological
Survey.

the Verdigris River. The story leaked out in Neodesha, but darkness fell before interested folk could go out and see for themselves. When curious groups proceeded to the spot the next day, there were no signs of oil. William G. Bryson, who was working with Mills, had gotten there first and repaired the damage.

Natural gas was still a phenomenon in 1893 when the Independence Gas Co. struck gas on the J.H. Brewster farm east of Independence. The company invited citizens to a display via the *South Kansas Tribune*. "The display was simply immense and exceeded in volume and brilliancy all expectations," wrote a reporter. "Through the timber pasture and about the residence, barnyard and in the road in front were a number of seething, roaring lights, 20 feet high and of sufficient brilliancy that one could read. There were derricks entwined with the national colors, and the sign of the gas company stretched between: 'Independence Gas Co.; C.L. Bloom, Pres.; A.P. McBride, Sec. and Treas.; J.D. Nickerson, Vice Pres.' The ample yard and the road in front swarmed with people, all in high praise of the display, and there was a long table set with rich cakes and lemonade and ice cream to tempt the hungry."

For Neodesha's Fourth of July celebration, gas was piped into town down Fourth Street and allowed to escape into the bottom of a barrel full of water set in another barrel. The two barrels were set in the ground about four feet. At night the spring was set on fire. It was a spectacular display of a force that had yet to be harnessed.

· D E E P   I N   T H E   H E A R T   O F   T E X A S ·

One of the first men interested in Texas geology was Elwin Theodore (E.T.) Dumble. "In the summer of 1888 when the legislature was considering a bill to establish the (Texas) survey I was called to Austin and had a conference with the Commissioner of Agriculture," he wrote. "The survey was a good beginning but...there was such a connection between mineral development and transportation that I presented my ideas on the subject to the management of the Southern Pacific Co. and was so fortunate as to enlist their cooperation and was appointed their Consulting Geologist." From then on, Dumble was active in management and petroleum exploration for subsidiary companies of Southern Pacific Railroad including Rio Bravo Oil, Kern County Trading and Oil Company, East Coast Oil Company and Associated Oil Company. Dumble felt that there was vast potential for the petroleum market. The "growing need for a better and cheaper source of locomotive fuel absorbed most of Mr. Dumble's time and attention," wrote a friend. Experiments suggested by Dumble and carried out by Southern Pacific's engineer proved that the crude oil was well suited for locomotives. It was a major step forward for the oil industry.

"Probably our most important service was in showing to the world the great benefits to be gained by applying geology to the production of oil," Dumble wrote. "In this the

Geological Dept. of the Southern Pacific was not only the pioneer in the region...but for some years we were the only company in which the work of oil development was controlled by the geologists."

Dumble was particularly interested in petroleum possibilities in Texas. "With youthful enthusiasm and the backing of Major J.W. Powell, at that time Director of the U.S. Geological Survey, I undertook in 1887 the task of trying to establish a State Geological Survey in Texas and a Chair of Geology in the University of Texas at Austin," Dumble wrote. Unfortunately, the plans went awry, and the person he wanted to fill the position was not approved by the legislature. Instead, Dumble was given letters from 125 applicants—"including men of every known profession except geologists"—and he was told to pick one. "I found the name of only one applicant who seemed to have the least idea of what the office was to be or who had received a post-Civil War college education," he wrote. That individual got the position.

In 1889, the *Dallas Morning News* published a two-column "Survey of the geology of Texas" that listed petroleum somewhere below guano, coal, metallic ores, mineral waters, gases and salt deposits. One sentence was devoted to oil: "Petroleum springs, including heavy lubricating oils, are known to eastern Texas and the work of studying this geological character will soon be begun."

Dumble required the field geologists with the Texas Geologic Survey to turn in written reports every 10 days. The result was, as one geologist put it, often half-baked, or worse, half-chewed reports and scientific results. Although the work lacked finesse, it went a long way in making political figures aware of the need and importance of geology.

There were problems, however, working under government strictures. Ralph S. Tarr, an assistant surveying in the Guadalupe Mountains, found himself caught between red tape and robbers when some of the survey horses were stolen. He reported the loss only to find the value of the horses deducted from his meager pay.

Texans seemed to put little stock in the survey, and promptly badmouthed it. Charles Gould wrote that trouble for the survey really started in 1892 when a "horny-handed son of toil" in the legislature claimed he had surprised a member of one of the field parties fishing. Determined to catch the culprit, he searched the auditor's books until he found an item of $0.25 spent for fish hooks. Like most of the citizens of the day, he did not realize that the field parties camped out and lived off the land as much as possible. Fish hooks were as logical an expense item as frying pans and axes. The legislator publicly took the survey to task for hiring wastrels. An angry public brought pressure to bear until the survey was officially terminated in 1894.

Surveyors pose at the Big Cedars at Fairy Creek Falls, British Columbia. The North American continent was covered with land which had not been explored in the mid-nineteenth century. American Heritage Center, University of Wyoming.

Oil had first been spotted by early hunters and trappers at Jackass Spring, Wyoming, 40 miles north of what would become Casper. A pool in the creek bed always sported a few inches of oil on its surface, and it never dried up, unlike the rest of the spring. Samuel Aughey, Wyoming territorial geologist, described it in 1884 as having at least 20 barrels of oil on its surface.

A young recorder with the geodetic survey of Wyoming, Montana and the Dakotas in 1890-92 noted the area which would later become known as the Williston Basin. "A well defined anticlinical fold extending from the Great Black Hills Uplift in Nebraska through northwestern South Dakota to the Missouri River....showing a prominent structure of immense proportions as yet unoutlined.....true pressure fold on the axis of which is exposed the Pierre Shale along the Moreau river....assuring oil-bearing horizons...muddy sands and Lakota....oil and gas for miles....Paleozoic limestone shows black oil." The men in the party were so impressed by indications in Southern Perkins County they nearly all purchased land in the vicinity of Bison, Faith, Isabella and Edson. "My husband bought 680 acres from weary settlers on Government land very cheaply," wrote Mrs. Adelaide V. Fisher in 1957. "We held (it) all these years paying taxes through leases to cattlemen always expecting oil would be developed." When her husband passed away, he reiterated to the family, "Hold that Dakota Land...Oil is there!" Like so many others—men who staked their lives on an idea—Fisher's faith remained steadfast. But the land remained untouched and unproductive.

Still further north, Eugene Coste, son of a well-to-do French family living in Canada, was making a name for himself in the oil business. Among geologists, that name was somewhat unprintable. Coste had worked with the Geological Survey of Canada for five years before deciding to explore for oil and gas on his own. He was convinced that oil originated from volcanic or Precambrian rocks. Like a few other lucky men, he managed to make a fortune despite his theory. His first big gas discovery was in Essex County, Ontario, on the shore of Lake Erie. It laid the foundation for Ontario's natural gas industry. Within just a few months, he discovered gas in the Niagara Falls area 200 miles away. In the early days when geology was still in its infancy, it was hard to argue with success even when it was based on erroneous theory.

A geologist poses beside a large boulder--a sample of glaciation. George Hansen Collection, Brigham Young University.

· S P I N D L E T O P ! ·

By 1895, Texas had produced less than 60 barrels of oil a year, "about enough to oil the guns of the Texas Rangers," a wag insisted. One man exploring Texas at that time was John Henry Galey. Galey was a wildcatter from the Pennsylvania fields. He was, his family noted, a prodigious student of geological literature. "He studied all sands meticulously and acted intelligently and scientifically on his findings," James A. Clark wrote. "He observed that gas was the greatest where the rock or coal formations were elevated the highest." From this he formed his ideas of anticlines. Galey was knowledgeable enough that Dr. William Battle Phillips, University of Texas geologist, sent Anthony Lucas to him when Lucas wanted help locating a well. Galey based his location on his interpretation of the topographic high being structural rather than adjacent to a sulphur spring. The result of Galey's theories was Spindletop, which literally blew in as the biggest well in oil history. With Spindletop, the craze for oil was cemented, and thousands rode, fought or walked west in search of black gold.

· B L A C K   G O L D   F E V E R ·

Just north of Texas, up in Indian country, several men skirted laws to drill on Indian land. In 1894, Michael Cudahy, of the meat-packing family from Omaha, began two wells at Muskogee. He must have missed the oil for he moved on to Bartlesville in 1897. There his luck changed only slightly. He struck oil all right, but since he had no way to transport it, the well was capped. A Territory cowboy by the name of George B. Keeler heard about oil discovered at Neodesha, Kansas. He had been interested in the black stuff ever since he had tried to get his horse to drink water from Sand Creek and had the discriminating animal refuse because of the scum of oil on the surface. He decided it was time to get a lease with the Cherokee Nation. Keeler joined forces with Cudahy and picked a site. When they tried to transport a string of tools from Tulsa, the roads were so bad that it took 14 teams to drag the equipment through the mud. The gamble paid off, however, in June 1897.

The Nellie Johnstone #1 was sunk to 1320 feet when the nitro shooter was sent for. "Mr. G.M. Petty, the expert shooter, arrived from Neodesha with the necessary material for shooting the well," wrote one eyewitness. "It took time to arrange the preliminaries, but everything was ready by 3:00 p.m., April 15. Miss Jennie Cass, stepdaughter of George B. Keeler, was allotted the task to drop the go-devil which exploded the cartridge in the hole. Very soon a large volume of oil, water and debris was spurting to the top of the derrick. Miss Cass has the distinction of shooting the first well in the Cherokee nation."

The Nellie Johnstone was claimed as the first well of commercial size in the Indian Territory. It produced a mere 50 barrels a day and was shortly shut down because of difficulties with the U.S. government and Indian leases. The next year, a bill was

Men pose at the gaging station, Hoffman's Ferry, Saratoga County, New York, in 1901. Note the early-day version of billboard advertising--a strategically painted barn. R.E. Horton, U.S. Geological Survey.

The Lance Creek well in Wyoming was located by C.J. Hares, north of Luck, Wyoming. Courtesy, E.K. Erickson.

introduced in the Territorial (Oklahoma) Legislature to establish a Geological and Natural History Survey of the Territory of Oklahoma. A.H. Van Vleet, professor of biology at the Territorial University, was given the title of territorial geologist.

·DIGGING FOR OIL·

Out in California, former gold prospectors and shoestring operators decided to dig for oil by pick and shovel. Plenty of men had already started exploring the Los Angeles city area when young mining engineer Edward L. Doheny happened upon the scene. Encouraged by surface indications in the Lakeshore Avenue area (later Glendale Boulevard), Doheny and Charles Canfield began hand digging a 4 x 6-feet shaft on a hillside only a few blocks from downtown Los Angeles. At 145 feet, the air in the hole became so laced with gas fumes that Doheny had to abandon his pick and shovel. They rigged up a spring pole drilling rig and sank another 15 feet of hole. At 160 feet, oil flowed in, filling the bottom of the hole. The Los Angeles field was opened.

News of the discovery brought a stampede to Los Angeles. Wooden derricks went up as fast as town lots could be leased or bought outright. When it became evident that the narrow structure ran almost due east-west for an undetermined distance, leasing agents tied up every available parcel of land as far west as what would become Western Avenue. The quiet palm-shaded community became a roistering oil field, as well after well came in.

Doheny and a myriad of others—Tom O'Donnell, Sam Cannon and Charles Canfield—made deals for leases on city lots and drilled wells so close together that legend persists that a string of tools lost in one well was fished out of another close by. Most of the oil hounds spent weekends loafing around the St. Elmo Hotel on north Main Street where proprietor L.A. "Daddy Ikey" Eichenhoffer, a German Jew, staked the oil boys to grub when they were broke or out of work. He was always paid back before anyone else.

Drilling became so frantic that ordinances were passed banning oil drilling near Sunset, Prospect, Hollenbeck and Westlake parks. Citizens who had been drilling the oil wells themselves took a different tack and suddenly began drilling "water" wells. Hundreds were drilled—and one did bring in a generous flow of hot mineral water where the famous Bimini Baths were later built. One field discovered by members of the Mormon Church was dubbed the Salt Lake field. More than half of California's total oil production came from fields within the Los Angeles city limits.

Even before the turn of the century, geologists were busy studying and mapping structures. In 1898, R.B. Lloyd and E.A. Rasor mapped the Ventura anticline; but they received no immediate response from sceptical prospectors who were hunting only faulted structures where seeps occurred. The fields, however, were structurally and stratigraphically complex. In 1902, G.H. Eldridge reported in the *U.S.G.S. Bulletin 213*, "The productive areas have been in every instance developed in connection with anticlines...In several instances faults, or intense disturbances of the strata, have accompanied the folding." Such geological structures required help from trained individuals—especially as obvious structures were tapped out and costs soared.

William Orcutt, an engineer, was hired by Union Oil of California to do both geological and engineering work. He worked from a spring wagon outfitted with provisions and tools. In 1902, he was working in the area of what would become the Santa Maria field. "At that time all small companies were fearful of being spied on by employees of the Standard Oil Trust," wrote R.P. McLaughlin. "He stayed in the local hotel and each morning he went into the hills armed with a shotgun and a game bag. In the evening, after carefully making his observations of the outcropping rocks, he delivered his game to the hotel cook."

·LIVING WITH THE NATIVES·

On the eastern side of the continent, one of the early geologists was W.J. Griswold. Griswold worked from Virginia to Ohio. He had begun his career as one of the original topographers with the U.S.G.S. when it was organized. In the hills of West Virginia, he and E.W. McCrary—as McCrary remembers—"lived with the natives. This back country work was hard, and living conditions were hard, but Griswold always took it in stride," he said. "He easily made friends with the people and adjusted himself to his surroundings."

But he was an intense man, according to McCrary. "I don't remember ever seeing him completely relaxed," he said. "His active mind was going from one idea to another generally along lines with his profession. Political and social questions were not of great interest to him—apparently. And oddly enough, he never seemed to be much interested in actively engaging in...production....He seemed to consider the production of oil as the business of some one else."

Griswold was convinced that his use of contour mapping would become a major tool for the petroleum geologist and for the companies who produced oil. "Unfortunately, however, only a small percentage of the men actually engaged in the production of oil attach much value to any geological theory," he wrote in an earlier time. "This is probably owing to the fact that the method of representing geologic structure by contours has not been previously applied to the oil-bearing strata, so that each oil operator might himself study the structural past. The many failures of those hunting oil on the anticlinal theory have thrown discredit upon the ability of geologists to assist in the location of productive territory. Most of these failures have been due to lack of knowledge concerning the geologic structure or to the absence of other conditions necessary for the accumulation of oil."

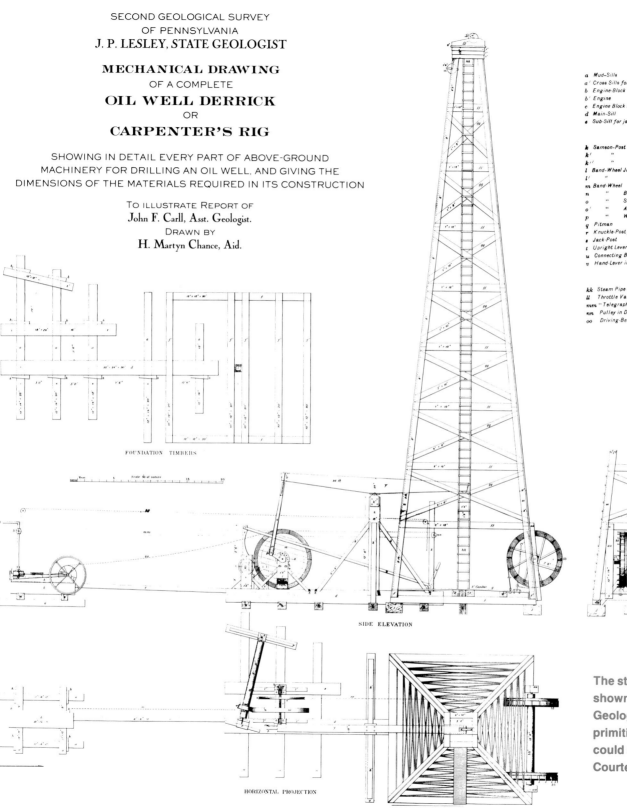
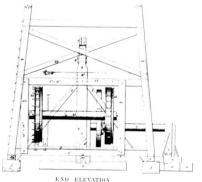

The standard oil derrick of the 1870s as shown in the Second Pennsylvania Geological Survey report may have been primitive by today's standards, but they could still drill for--and find--oil. Courtesy, Jeffrey Heyer.

''The advice a geologist can give is much like that of a surgeon. He draws on his total knowledge, experience and the facts he has, says a prayer if he is a religious man, and then gives his best judgment. If he is a good geologist, he is right more often than he is wrong. But the proof whether he is right or wrong comes only when the oil is found or not found. Like medicine, geology has developed many ingenious ways of scanning the subsurface of the earth, such as sending down seismic shocks and deducing the density of the rocks by the echo. But after all this 'X-raying' and laboratory work, you still must drill—as the surgeon must use his scalpel—to know for sure if your surmise was correct.''

Lewis Weeks

(Preceding page)
The oil strike in Bartlesville, Oklahoma, north of Tulsa sent men rushing into Indian Territory to seek their fortune. The photo shows Bartlesville in 1903.   Phillips Petroleum.

(Right)
A high peak, such as Garfield Peak in Natrona County, Wyoming, offered the geologist an excellent triangulation station.   C.J. Hares, U.S. Geological Survey.

(Opposite)
Oil exchanges were set up in the Pennsylvania area for men who relished gambling and tried outguessing crude prices. The Oil City Curbstone Exchange posed in 1870. Many men lost fortunes when the oil exchanges crashed in the 1880s.   McLaurin, Sketches.

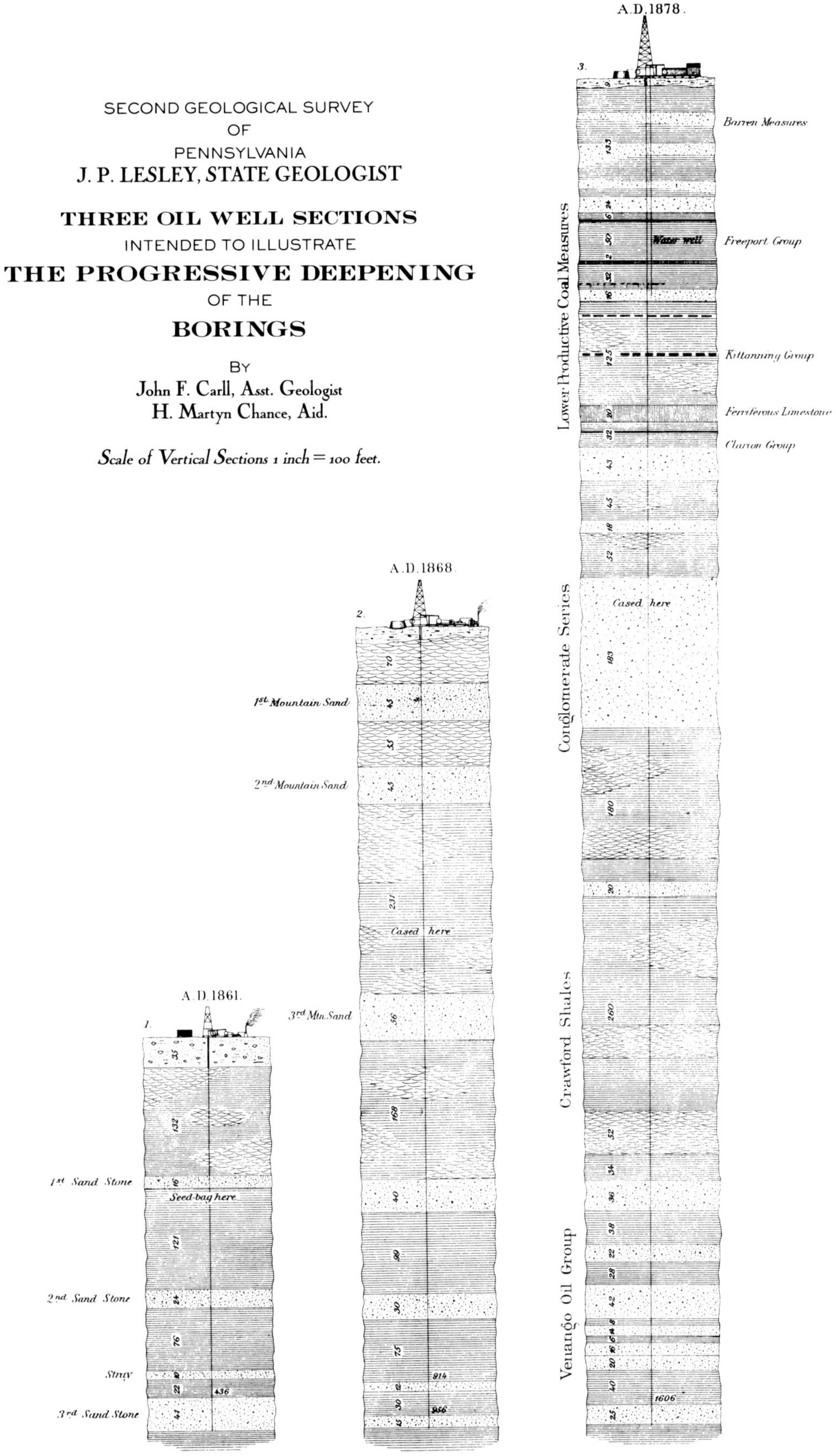

As technology improved, drillers began to go further into the earth, and the need for an understanding of deeper structures and sands became more apparent. The Second Pennsylvania Geological Survey compared depth changes that had occurred in only a few short years. Courtesy, Jeffrey Heyer.

(Opposite)
Geologists who worked for the U. S. Geological Survey found the travel primitive. Transportation across rivers--such as the Broad River in Madison County, Georgia--was often by hand-poled ferry. M.R. Hall, U.S. Geological Survey.

SECOND GEOLOGICAL SURVEY
OF
PENNSYLVANIA
J. P. LESLEY, STATE GEOLOGIST

THREE OIL WELL SECTIONS
INTENDED TO ILLUSTRATE
THE PROGRESSIVE DEEPENING
OF THE
BORINGS
BY
John F. Carll, Asst. Geologist
H. Martyn Chance, Aid.

Scale of Vertical Sections 1 inch = 100 feet.

·AMBASSADOR FOR A DAY·

Albert Enrique Fowks cut his teeth roaming the Pennsylvania oil fields on a burro, watching the drilling crews at work. In 1878, he and his mother and brother joined his father in the Peruvian oil fields. When he arrived, he recalled that the landscape was barren and uninviting, nothing but a few crudely housed natives, struggling for existence in a rigorous environment. The hacienda "La Brea de Parinas," on which the first Negritos wells had been drilled, was an old Spanish grant. In 1888, H.W.C. Tweddle, who had discovered and developed the Baku oil fields of Russia, purchased the grant. Tweddle and Englishman William Keswick organized the London and Pacific Oil Company to exploit the hacienda. Fowks and his father remained on the property as advisors. Albert was then sent to Payta for the new organization. On September 20, 1893, he married May Goddard, daughter of an American resident. The ceremony was in the American Legation in Lima—the only place in Peru where the nuptials could be performed under the law, since they were not Catholic and could not be married in the churches.

As it happened, the minister was to leave for vacation immediately after the ceremony, and he graciously turned his quarters over to the newlyweds. During the three weeks they stayed in the Legation, they were accorded all honors normally given to accredited representatives of the United States. That included being solemnly saluted by the Legation guards each time they appeared outside the building.

Fowks and his wife returned to Payta at a time when Peruvian petroleum developments were commanding world-wide attention. He was commissioned by J.C. McDowell, vice president of Standard Oil, to provide periodic progress reports from the area.

About 1900, Chester Brown organized the Titicaca Oil Company, and Fowks became involved. The property was around Lake Titicaca, the highest body of navigable water in the world. The first five wells brought in small yields of high-gravity crude. The fourth well came in out of control, producing 1,000 barrels in the first six hours but dropping away later to 150 barrels a day. Much of the big flow was lost because there were no storage facilities. The field, however, was too far removed from tidewater to be operated with any degree of comfort or profit and was later sold.

Women often moved their families close to drilling sites just to be near their husbands. Many a wife complained when the well came in, because it ruined her wash. Note that the top of a hill was still a preferred place for drilling, because of the belief in "drilling on the highs." National Archives.

·DOWN MEXICO WAY·

As early as 1889, Elwin T. Dumble and Josiah Owen decided that large oil pools lay beneath Mexico. Dumble tried to interest Southern Pacific, but it was several years later—after the president of the Mexican Senate requested Dumble's help—before things began to pan out. Dumble was sent as consulting geologist for Southern Pacific to select lands for the company's newly formed East Coast Oil Company.

At that time, Tampico was a fever-ridden malarial village beside a beautiful river which was reached by a branch of the Mexican Central Railway. A.A. Robinson, president of Mexican Central, invited Edward L. Doheny and Charles A. Canfield to investigate the Tampico area. Doheny had prospected for gold in Mexico 20 years earlier as a young mule-driver for the Geological Survey in Arizona and New Mexico. They liked what they saw and bought up 450,000 acres west of Tampico, and 170,000 acres south toward Tuxpam, for 60 cents an acre—an astronomical sum to the Mexican owners who thought it only a worthless jungle. At first, Canfield and Doheny cut and crawled their way through the jungle themselves, hunting for signs of oil. Finally they gave up and offered five pesos to anyone pointing out tar spots or seeps. Robinson provided Doheny with a special train for prospecting oil seepages in the Tampico region along the railway line. Every few miles, Doheny would stop the train while he and his assistants explored the fields on foot. It was rough, harsh country, sparsely populated, with a few scattered rancheros or an occasional hacienda where some lordly inheritor of a Spanish land grant lived in baronial splendor.

Their prospecting paid off, and Mexican Petroleum of California opened the "northern fields."

In 1901, W.H. Dalton, an English geologist, investigated Mexico's oil prospects for S. Pearson & Son interests. In 1902, H.B. Goodrich, a consultant for the Doheny interests, made the first commercial geological report on the Tampico district. London Oil Trust completed a small well at Cerro Chapopotal in the "southern fields," and the Pearson interests began drilling on the Isthmus of Tehuantepec. Geoffrey Jeffreys began work in 1904 for Pearson. Jeffreys was an Englishman, a rugged character equally adapted to life in the jungle or in formal diplomatic and financial circles. He walked over much of the country from the Isthmus of Tehuantepec to Sierra Tamaulipas, describing rocks and measuring sections.

The first Mexican well of economic importance was Mexican Eagle Oil Company's No. 1 Pez, near Cerro La Pez in the Ebano district. Jeffreys was on the Dos Bocas discovery well which blew in July 4, 1908. It immediately ignited and furnished some of the most spectacular fireworks in oil history. The hole cratered and began flowing salt water August 30, but it signaled the beginning of an extensive search south of Tampico in what became known as the Golden Lane.

Everette Lee DeGolyer was an early recruit of Lord Cowdray (Sir Weetman Pearson). DeGolyer was joined by his friends, Edwin B. Hopkins and Chester Washburne. When DeGolyer arrived in Tampico, he put up at the new Victoria Hotel, run by an American widow, Mrs. Weeks. One morning, DeGolyer and Hopkins set out by launch for the Tamiahua Lagoon below Tampico. Superintendent Calder went down early in the morning to see them off and informed them that if the launch broke down (a common occurrence), they were to open up the emergency rations on board. The launch did break down, but they were able to pull into a little estuary where friendly natives treated them to beans, tortillas and coffee. They never had to break into the emergency rations. However, when they returned, DeGolyer's curiosity got the better of him, and he investigated the groceries. One case contained tomato catsup, another Lea and Perrin's Sauce, the third bottled vinegar.

Most of the geological work in Mexico was done in the back country, on horse or mule. "The geologist did not operate from camps, as the drillers did; he was expected to go out alone, to survive, to live off the country," DeGolyer once noted. The first automobile had not yet arrived in the area, and there were few roads. Prospecting had to be done on horseback. If the horse was stolen—and it often was—the geologist turned to the burro or the mule carrying the saddlebags and the tent for the pitched camp. "The young American geologists had an advantage over many of the English and European scientists," DeGolyer explained. "The Americans had ridden horses since childhood and had usually dealt with the maddening stubbornness of mules. The Texas saddle was in general use, and the Americans were accustomed to it."

The geologist usually traveled light—a hand compass, hand level, a lens and geologist's pick or hammer. A few lucky ones, like DeGolyer, had a plane table. Dr. J. Th. Erb, well-known Swiss geologist, was envied for his pocket range finder.

The geologist slept in a tent and was his own cook. Near the sea, food was no problem because Gulf oysters could be scooped up along the beach. Inland, hamlets observed the ancient tradition of hospitality to the stranger. Food at the ranches ranged from shrimp in banana leaves to tortillas, black beans, dried meat and eggs. Furthermore, a geologist could find sustenance at any company's camp. There he could count on food and a bunk. Naturally, he would be plied with questions, but a good man knew how to eat with his mouth closed.

Since field work could be strenuous, geologists often found themselves creeping along riverbanks that bristled with thorny plants. By the end of the day, it was not uncommon for a geologist's clothing to be reduced to tatters. The standard outfit in the tropics was a cotton shirt, riding breeches, knee boots and a stiff-brim hat with fine mesh netting which hung down over the shoulders. Despite the netting, gnats always managed to get through; and by the end of the day, a traveler's face and neck were freckled with tiny blood clots from the bites. In such areas as the Isthmus of Tehuantepec, it was so hot and humid that the men found it impossible to wear protective gloves or even a head cloth. To compensate for their exposure, they fortified themselves against the inevitable malaria with a daily potion of quinine and wine.

Another creature that plagued the geologists was the tick. After some experimentation, the men discovered that chewing gum was the best way to remove an active insect. However, if it had already gotten into the skin, a touch of tobacco leaf soaked in *aguardiente* usually managed to make it turn loose. Another effective means for removing the offending creature, according to Robert H. Dott, was to hold a lighted cigarette to its rear end.

Seepages were still the standard method of spotting promising locations, and once the natives caught on, many requested money before they would divulge locations.

Before advanced printing techniques,
photographs and drawings were
translated for newspapers through the
engraving process. The etching is
described as "Hot Springs on the
Gardiner River on the Upper Basin."
Blakey Group Collection.

Despite the hardships, the work paid off, and Mexican Eagle hit it big when DeGolyer picked the location of the spectacular Potrero del Llano #4, which produced a flow of 110,000 barrels a day—a record at the time.

Not everything was rosy, however. When one well caught fire, Lord Cowdray wrote DeGolyer, "The men will not be paid today. Advise them that the authorities tell us the men must remain at work until the fire is extinguished and that we must not pay them till then. Otherwise the men will leave. They can have some payment on account but we must hold the regular pay back....Make the best terms possible with the woman feeding the men. Fifty cents a day for 3 meals per man should be a fair charge under the circumstances. But do the best possible, threatening that you will ask the *rurales* (local policemen) to see that she feeds the men and have the local jefe or judge tell us what we must pay her. This only if she be too unreasonable."

·PIPELINES AND PITS·

Oil was everywhere. The Baku district on the Apsheron Peninsula supplied 95 percent of Russia's total output which for many years surpassed U.S. production. A pipeline from Baku to ports on the Black Sea improved access to the world markets, but strikes and political disturbances reduced production in the entire area. The Baku was a messy field. Uncontrollable gushers blew oil over everything. Ordinary wells put out tons of sand along with the oil. Natural gas was a menace. Since there was no market, millions of barrels of distillate went up in deliberate flames. Many small operators were insufficiently financed and inadequately staffed to handle the problems, and even such companies as Nobel and Rothschild were hard-pressed to handle the problems.

In Borneo, Royal Dutch geologists struggled through the rain forests. When they cleared a path, they dug a pit 15-25 feet deep through weathered mantle down to bedrock. The pit was wide enough for the geologist to crawl into, and from these holes, he measured the dip and strike of the strata. A rough survey was run to locate the pits, and a sketch map was drawn showing the topography and structural observations. Labor was cheap, and the company employed young Dutchmen, who had been raised in the Indies and knew the native language, to supervise the work crews. The geologist had little to do but guard himself against jungle creatures, pit cave-ins and unhappy workers—and find oil in the process.

·TABOO TERRITORY·

In 1900, Charles Gould, an energetic young professor at the Territorial (later State) University of Oklahoma, was named geologist for the territorial survey. Gould's department at the university had no facilities, and he was the only employee. What geological library and collections there were belonged to Gould, personally. He worked at a desk in a shared office and taught in a borrowed classroom. Despite such limitations, he offered eight courses that first year—elementary, advanced, and economic geology; mineralogy, and physiography as well as commercial geography, geological biology, and paleontology. One of his major contributions was the establishment of field trips. Gould took students to the Arbuckle Mountains where they could examine geology firsthand.

In October 1901, the first gas well of commercial size was discovered in a residential part of Tulsa, Indian Territory, between Seventh and Eighth streets and Boulder and Cheyenne avenues. Although it was only a shallow gasser and was never used commercially, it created interest in the oil and gas possibilities in the area. It was shut in and blown off for the benefit of Pennsylvania oil men on their first trips into the territory.

Despite the strike, a report by the Department of Geology and Natural History of the neighboring Oklahoma Territory in 1902, was less than enthusiastic. "From its geological position, oil and gas in Oklahoma (Territory) must be at a considerable depth....While it would be unwise, in view of the many surprises of late years in the discovery of these products, to predict the presence or absence of them in any locality, the depth of the oil-bearing strata may be estimated with considerable certainty, and this knowledge, if made use of, would prevent much waste of money in foolish experimenting."

Few geologists bothered to investigate the future state in the early years. The problems of Indian leasing, as well as poor or nonexistent transportation, hampered the industry. H.B. Goodrich, consulting geologist for the Santa Fe Railroad, was one of the first in the state. He moved to Ardmore in 1903 to work for Coline Oil, a Santa Fe subsidiary. When restrictions on leasing of "unproved lands" were lifted in the (Indian) territory in 1904, production took off. In 1905, Goodrich brought in the Wheeler gas field, northwest of Ardmore, Indian Territory. Up around Tulsa, in the area known as Glenn Pool, Bob Galbreath and Frank Chessley had scouted the area with horse and buggy and proceeded to bring in the discovery well in November.

·MUD SMELLERS AND TRAVELIN' MEN·

The strikes in Indian Territory and others in Texas created a flurry of activity. Texas was besieged with oil men, land men, geologists and hucksters—and it was often hard to tell the difference. For the next five years, fields sprouted throughout the state. Most independents ignored the new science—geology—that seemed to show few immediate results. They were,

Charles N. Gould, professor at the University of Oklahoma, trained many of the young men who went on to become well-known geologists. Courtesy, Donald Gould.

after all, interested in the bottom line—and that meant oil at the bottom of the well. In the earlier days of what he termed the scientific approach, "the attitude of the 'practical man', so-called, was almost entirely hostile," John C. Woolnough wrote. "The driller resented the intrusion of the 'mud smeller,' as he contemptuously called the geologist, into what he considered his own particular province, and, not infrequently, deliberately set to work to confuse and injure him by supplying false information and samples."

Despite such hostility, geological surveys expanded throughout the country. They drew heavily upon young men from colleges and universities for their manpower. Even professors, such as Charles Gould, worked for the U.S.G.S. during summers. His summer field reports had a direct bearing on the development of many of the early fields and called attention to anticlines in southeast Oklahoma, oil-sand outcrops in the Arbuckle Mountains and gas seeps in eastern Oklahoma. A spot on one of the field surveys was considered a plum.

Charles W. Hamilton had no money when he enrolled at the University of Oklahoma. He heard that the Oklahoma Geological Survey was going to put a field party into northeastern Oklahoma to map surface formations, so he lined up with the other students to apply. Although he had no geological training, he begged Charles Gould and Dr. D.W. Ohern to give him a job. Others, he was told, were better qualified. He persisted. Gould lauded him for his persistence but finally told him that all regular field jobs were filled. There was only one opening left in the party. "Can you cook?" Gould asked.

Hamilton's mother had been afflicted with rheumatism and had taught him to help her in the kitchen. When he replied that he could turn out pies, cakes, bread and other sundry culinary dishes, he was immediately hired. He cooked for two months before graduating to rodman. The trip changed his life, and he signed up as a geology major.

The parties on the summer trips managed to live off the land and what supplies they could load in a covered wagon. On one of Gould's trips, supplies ran low and he went into Dougherty, Oklahoma, to replenish their stock. There was only one store in town—run by a German immigrant. Gould called for what he thought was a simple order—two loaves of bread and a pound of bacon. The German replied that there was no "baked bread for to sell;" women in Dougherty all baked their own biscuits. Gould then asked to purchase a small sack of flour and was informed by the merchant that the last sack had been sold the day before and no more was expected in town until the following week. Gould tried again—this time calling for a box of crackers. The merchant replied stoically that there were no crackers in town because no one bought them. "I'll tell you what, my friend," he told Gould, "in this town it ain't what you want, it's what you can get."

At the University of Wisconsin, C.K. Leith also used the field party technique to enamor young men with the wonders of geology. His first field parties were into northern Wisconsin, nearby states and Canada to locate iron ore and other metals. The party usually consisted of one or two geologists, a mineralogist, several compassmen, packers and a cook.

It was not long before the *Outcrop*, an irreverent annual account of camp life, mysteriously appeared. Early volumes of this epistle documented the rigors of field work—the countless cedar swamps, black flies, "no-see-ums" and endless portages. After C.K. Leith was treed by a moose on one of his visits, a piece appeared entitled, "Moose I have known." In 1904, an appendix to volume 2 included a "Special Dissenting Minority Report" by those non-geologist members—packers, cooks and compassmen—upon whose shoulders the burden of daily life fell. The report outlined a number of now generally accepted fundamental geological truths, such as "A portage is the longest distance between two points" and "There is a linear and exasperating relationship between occurrences of iron formations and great swarms of black flies and mosquitoes."

Despite such discomforts, many young men were drawn to the prospect of travel. "Nobody asked me to be a geologist," Hugh Miser explained. "My mother wished me to be a preacher. My father wished me to be a politician. I became neither. I wished to be a civil engineer who makes maps and travels. Father responded, 'A civil engineer is a traveling man; a traveling man is no good. I will not have a no-good-traveling son'....At college I learned that a geologist makes maps and travels; I decided to be one then and there. Of course, father's attention was not immediately called to the matter of travel."

Charles Gould (center, bearded), was
responsible for the interest of many
young men in geology. The geological
party on the Verdigris River in 1909
included D. W. Ohern, Everett Carpenter,
Artie Reeds, Ben Belt and Bob Wood.
Western History Collections, University of
Oklahoma.

(Above and opposite)
The U.S. Geological Survey moved into the Yukon region of Alaska in 1903. Young men try training a horse to pack, at Eagle district. The horse often came out on top, much to the chagrin of the greenhorns. L.M. Prindle, U.S. Geological Survey.

Cooks at the Eagle U.S.G.S. camp
prepared such delicacies as beans, bacon
and cornbread. Fish came from the river,
and--if they were lucky--some four-footed
critter might wander close to camp and
stay as dinner. Note that there are several
rifles--not just one for the cook. Each man
was expected to do his part of the
hunting, and every rifle was needed to
protect the mules in "bear country."
L.M. Prindle, U.S. Geological Survey.

Smuggler's Canyon, Brewster County, Texas, was typical of the remote areas to which geologists were attracted. The photo was taken in July, 1904. Barker Texas History Center, University of Texas at Austin.

·No Geologists Wanted·

About 1905, William E. Wrather was a young undergraduate at the University of Chicago taking courses toward a law degree. "I fell under the influence of R.D. Salisbury," he said. "He decided I had some talent as a geologist and influenced me to take a geological course. Nevertheless, I did not give up the idea of law. While in school I took a spell of typhoid fever and almost died. The doctor recommended outdoor exercise, so a geological job seemed logical.

"W.H. Emmons was a graduate student in the University and he interceded with Dr. Hayes of the U.S.G.S. I went on the payroll at $60 a month. I had one of the most glorious summers I ever spent in my life.

"That fall and the following year I took law courses and became thoroughly convinced that I was cut out for a geologist rather than a lawyer."

When Wrather graduated from college, he discovered to his dismay that the U.S. Geological Survey where he had been promised a job had suddenly had a slash in funds. "I was left high and dry with about two or three hundred dollars in my pocket, with no job," he said. "Oklahoma had just made it as a State, and I took a little of my remaining cash and beat it for Oklahoma." The oil business was going strong in 1907, and Harry Hanson put him in touch with F.A. Leovy, head of Gulf's Oklahoma division. Leovy offered him a job as a scout in Beaumont, Texas, with the Guffey Co., then a part of Gulf. "I wasn't keen about going to Texas because I'd recently had a spell of malaria and that's malaria country," Wrather said. But Leovy replied, "Well, you can take the job. You don't have to stay if you don't want to. I lived in Beaumont for a number of years and never died, so you might try it." Wrather took it.

"An oil scout is a general handyman," Wrather explained. "He's an information man who watches wildcat wells, reports on showings, finds where leases should be taken, and keeps tab on everything that's happening in the neighborhood in the oil business." (Mr. Markham) knew I had had geological training and said: 'Now young fellow, I'm not hiring you as a geologist. You're a scout. You can use geology if you want to, but the oil companies don't have much to do with geology. They don't have any confidence in it. And if you could use your geology, that's all right, that suits me fine, go ahead. But you're not employed as a geologist.' "

·Too Damn Many Questions·

Wells were sprouting and spouting all across Oklahoma—Red Fork, Glenn Pool, Bartlesville, Dewey, Chelsea. In 1906, W. Dow Hamm's parents moved to Indian Territory. "I had been afflicted with asthma at Bentonville (Arkansas)," he explained, "but on arriving in Muskogee the aroma of the sweet shallow oil production in the old Muskogee field proved to be an immediate cure and, in fact, oil still smells good to me.''

Most operators in Oklahoma relied on their ability to 'smell' oil. If they were not adept at that, they sought out an undiscovered oil seep or drilled near someone else's success. They thought little about geologists one way or another. Some looked upon the sudden crop of educated youngsters as dandies. One driller refused to let a geologist on the derrick floor because he wore one of those ''fancy new'' wristwatches. Others saw them as competition. A geologist who approached a well near Dale backed off when he spotted a shallow grave with headboard and footboard near the well. At the foot of the grave were the toes of a pair of boots sticking up; on the headstone was scrawled, ''He asked too damn many questions.''

·SHAKEDOWN·

Many of the young geologists in the California fields came from Stanford University, where a strong geological school had been introduced. In 1906, Harry Roland Johnson was enrolled in Stanford University. A member of Delta Upsilon fraternity, he was also something of a painter and had polished off an advertisement for a Greek tragedy to be presented by Stanford's Drama Club. The morning of the San Francisco earthquake, Johnson was shaken out of his room. ''When the shaking was over we went down to the Quad to inspect the ruins,'' wrote his roommate. ''There was the great arch at the entrance to the Quad with about 50 feet of the right corner broken off and great cracks extending down through the structure, and tons of huge rocks from the frieze of carved figures at the top of the arch lying on the ground. Harry said, 'I feel queer inside,' for right there before our eyes was an exact replica of (the poster)....In the weeks to follow, Harry (walked) the surface trace of the San Andreas fault from back of the campus to San Francisco, measuring offsets in roads, fences and the vertical displacements along it. Here was something made to order for him and all hell could not have kept him from looking for the causes of the slip.''

In 1907, five new Stanford graduates were hired by Kern Trading, the California subsidiary of Southern Pacific, to serve as ''resident geologists'' under F.M. Anderson. They were the first ''well sitters'' ever employed on a regular basis in the oil industry. They were responsible for maintaining subsurface maps, picking casing points and advising on sand penetration.

About the same time, interest was growing in the northern United States, particularly in Wyoming. Cesare Porro, a noted Italian geologist working for Petroleum Maatschappij (Company), investigated the Salt Creek area. ''Salt Creek started out as a swindle,'' Silas Lane explained. Lane was working for Cecil Rhoads in Africa when Lord Templeton called him to go to Salt Creek. Lane asked where Wyoming was located. Lord Templeton—as ignorant about it as Lane—told him to find out. Lane asked why there was so much sudden interest in Wyoming and Salt Creek. Lord Templeton informed him that he and his associates were "a bit suspicious" that they had been sold "a bit of worthless land. We don't accuse anyone of fraud, we are only suspicious," Lord Templeton insisted. Lane found Salt Creek and set up a drilling test which erupted into a spectacular well on October 23, 1908.

From there the fever spread across Wyoming—to Carter, Dallas, Shannon, Greybull (a gasser) and later Grass Creek.

Henry Sherard was an early geologist in the area, according to C. J. "Charlie" Hares. He had picked up his knowledge of fossils, formations and anticlines while driving the elder Knight, of the Wyoming School of Mines, in his geologic work all over the state. "Sherard was an inveterate cigarette smoker, he simply had to light one every time we opened a barbed wire gate," Hares said. "Once his cigarette ashes caught my pants on fire to darn near burn me up. 'By golly' was Sherard's favorite cuss word."

·BY GUESS AND BY GOD·

Getting recognition was still a difficult task, on the whole. "During this period, geology was only beginning to be recognized as being of value in the discovery of oil," wrote Elizabeth A. Ham in "A History of the Oklahoma Geological Survey 1908-1983." "Oil was found by guess and by God, by instinct, by smell, by a feeling in the bones, by doodlebuggers, by luck or by unidentified skill—not by identifying structural or stratigraphic traps, not by geologists."

Geologists, however, were determined to turn that around. In 1908, the Oklahoma Geological Survey was established, and Charles Gould was named director. Within an hour after his appointment, he had arranged for five geologists to begin field work immediately. By summer, he had nine parties in the field investigating oil and gas fields in Tulsa, Creek, Okmulgee, Muskogee and Wagoner counties.

That same year, California's Associated Oil Company, which had been taken over by Southern Pacific, formed its own geological department. Other companies began to follow suit. But such procedures would not be adopted in the Mid-Continent region for another decade.

·WHISKEY ROW·

Dennis Leo Driscoll arrived in Coalinga, California, on a Sunday evening in February 1908. He was hungry and tired after his long journey from Bradford, Pennsylvania. The last part of the route from Hanford to Coalinga over a branch line of the Southern Pacific was a nightmare of discomfort. To top it all, it was a wet, cold, miserable day when he stepped from the train. He started walking. In a few minutes, he arrived at Whiskey Row. The sidewalk was about three feet above the ooze that constituted the street and gangs of oil

Geology students from the University of
Oklahoma unload their gear in 1909. The
party included young women as well as
men. This probably was from a special
train at Dougherty on the Santa Fe R.R.,
in Oklahoma's Arbuckle Mountains.
Western History Collections, University
of Oklahoma.

men were having fun trying to push each other over the edge into the morass. Every time one took a dive a roar of laughter went up from the crowd that packed the block almost solid. Driscoll repaired to a quiet corner and phoned his new headquarters for help. A buggy was sent and he was driven nine miles to the camp where a nice cottage awaited him.

Driscoll soon grew accustomed to Coalinga's character. Its main attraction was Whiskey Row, a solid block of saloons that never closed down. "Every establishment in the block had several tables of poker and blackjack constantly in progress," he said. "The dance halls were across the tracks directly opposite The Row and they, too, ran wildly and uninterruptedly. The City Marshall mingled with the crowds but had little to do as no one ever stepped out of line. The men were boisterous, but always law abiding. It amazed me to see the stacks of twenty, ten and five-dollar gold pieces piled up like poker chips in front of the players, and it wasn't uncommon to see a whole stack of these gold pieces shoved to the middle of the table on a single bet. We rarely saw a gold piece in the east, and this affluence fairly made my eyes pop. When I left Pennsylvania we were being paid our wages in scrip and any kind of money was as scarce as hen's teeth."

Despite the wide-open town, there was little lawlessnesss in Coalinga. "Women were quite safe on the streets at any time of the day or night—even the dance-hall girls—and anyone who dared violate the unwritten code by which they were protected was flirting with life itself. There were few trees for the accommodation of lynching parties, but one of the timbers on the railway bridge just out of town had an extension that was reputed to be ideal for such a purpose.

"Men trusted each other. Nobody ever locked a door. You could make a deal with anyone, stranger or otherwise, secure in the knowledge that the terms would be rigidly adhered to."

· A LITTLE NIGHT MUSIC ·

The Taft-Maricopa area was much the same. J.R. "Bill" Pemberton worked for Edward L. Doheny's Pan American Oil Company and was assigned to the area in 1908. In those days, oil land in California was obtained under the old mining laws. Claims were 'staked' on a first-come, first-served basis. Daily scenes were enacted which were much like those in the earlier mining days in the west. Horse-drawn drays loaded timber for rig construction and headed out of town in search of open ground. Claim-jumping was common, and it was not unusual for a prospector to return from filing his claim in town to find someone building a derrick on his land. The geologist was really a scout whose quick response could make or break a fortune. He searched out information on who was drilling where and what they were finding in any feasible manner.

Pemberton was known for his amazingly detailed reports which he called in to the Los Angeles office every morning at 9 o'clock sharp. The reports were one of the reasons that

Soda crackers and canned peas were typical of the fare in the mess gear in 1907. This particular mess box was photographed in Wheatland County, Montana. The photo also shows preserves, canned tomatoes, and a clock for timing bread. This crew ate pretty well. R.W. Stone, U.S. Geological Survey.

Anna Anderson #1 West Columbia oil field well doing about 800 barrels. Trust-from Mrs. Hugh Hallburg

Geology students listen as D. W. Ohern expostulates in the Arbuckle Mountains, in 1909. Western History Collections, University of Oklahoma.

Brothers Frank and L.E. Phillips were nearly destitute after three unsuccessful wells. Then the Anna Anderson No. 1 gushed in September 6, 1905 and was the first of 81 consecutive successful strikes. There was no geological speculation here--the men simply drilled as close as possible to a producing well. Phillips Petroleum.

Doheny was so successful in the field. But no one knew how Pemberton came by such valuable information. Some years later, someone asked him how he managed to accumulate so much accurate information on a daily basis.

"Daily!" Pemberton scoffed. "My work was all done at night. Instead of rushing around all over the place like all the other scouts, I simply got a job playing piano in the local whore house. If I didn't get all the information I wanted from the clients, who were the best drillers in town, I'd fill in the gaps from the girls."

By 1909, California led the world's oil-producing regions. Excitement was so high among the general public that fakirs were locating oil claims for trusting but naive customers in the Tulare Lake area. When some of the more gullible traveled out to see what their $5 or $20 contribution per location payment had bought, they found their parcels under water. Pennsylvania oil men thought the California oil companies were in the same class of character when they read their reports. Oil could not possibly flow through the pipes as fast as many of them claimed.

·CONTOURS AND CONVERSION SHEETS·

In 1909, Kessack Duke White was a sophomore at the University of Kentucky. "I had a summer job with the Kentucky Geological Survey as instrument man and assistant to Fred Hutchinson, whose assigned problem was to map structure and evaluate petroleum prospects in Western Kentucky," White wrote. "In 1910, I received in April of the spring quarter, from Dr. White, president of the University, permission to leave classes to accept a job with Hutchinson and McCrary for structure mapping in Ohio and West Virginia.

"The method used was to trace beds of coal or thin limestone determining the location and elevation at outcrop points. The elevations of localities, other than the Pittsburg coal outcrops, were converted to the Pittsburg coal by adding or subtracting the known measured interval. The Pittsburg coal was used as the key horizon for the structure contours. Where not feasible another horizon was used, such as the Ames limestone in Ohio, in the same manner.

"In order to project the surface contour map to depth an isopach map was constructed on tracing paper to serve as an overlay convergence sheet. Data for the isopach map were obtained from wells drilled in the area, or if absent, from wells drilled in surrounding areas. Thus, by subtracting the interval from the data of the contoured key horizon, a subsurface map of a bed, such as an oil sand, could be constructed from the key horizon structure map, with reasonable accuracy by compensating for variation in intervals.

"The history of the use of contours for structure delineation began with the Rodger brothers, of the Pennsylvania Geological Survey. Then it was somewhat controversial....The U.S. Geological Survey assigned (M.J.) Munn and (W.J.)

Geological surveyors pack their gear at the end of the season in 1910 and head back on their way from Harney Creek to Walker Creek, Converse County, Wyoming. D.E. Winchester, U.S. Geological Survey.

Griswold to test and prove applicability and feasibility of using contours and conversion sheets for the construction of geological structure maps and their use in petroleum exploration and development.

"At that time two hypotheses were current in Appalachia: The 'anticlinal,' applied by Prof. I.C. White, and the 'Trend,' generally followed by oil field personnel. Both, of course, were applicable."

The field party consisted of a "geologist, carrying a (Wye) level and a rodman carrying a level rod. Location was visual using a U.S.G.S. quadrangle map which had been blown up 4×. It was quite accurate....

"Our time was fully occupied and our expense allowance quite limited. As a result, in our headquarters town or village, we rented a room with a private family. Monday morning we left, not returning until early Saturday afternoon. Sunday was spent in our

The tent at Valdez district, Alaska Gulf
region, included stove, bedrolls,
equipment and one sleeping surveyor.
F.C. Schrader, U.S. Geological Survey.

room checking notes and maps. We covered the country on foot, stopping at night, for bed and meals, at a nearby farm house. This took care of supper and breakfast. Lunch was what was available at cross-roads stores.

"In walking the outcrop we would generally go up one side of a valley and return by the other side....When going up the valley (we) would tack 3-inch squares of white cloth to stragetically located trees so that when crossing the valley to walk the outcrop on the other side we would have our elevations established without having to carry the line."

White's summer work "started an exodus of students trying to earn summer money," he wrote. "In those days a large number of families still felt the pinch of poverty, created and left over from the Civil War. So students helped out as best they might with earning what few dollars they could."

After graduation, White joined the Kentucky Geological Survey, working with James H. Gardner and Julius Fohs. "Gardner was in coal, Fohs in Fluorite and I in oil. Fohs was a rather unusual person. He lived in a one-and-a-half-story cottage in which a room under the eves (sic) was his study. He worked full time writing his bulletins in this little room. He was an upright, conscientious, honest, honorable individual.

"At the beginning of 1912, Dr. Gardner obtained for me an assistant geologist place on the Illinois Geological Survey through his friend and earlier associate, Dr. Frank DeWolf, the director. Years later Dr. Gardner told me the reason he got me the job was to send me north so that I would learn that 'Dammed Yankee' was two words."

There were others mapping oil-producing areas, particularly for the surveys in Pennsylvania, West Virginia, Ohio, Indiana and Illinois. All published reports on the oil fields of their states. Only a few operators followed their recommendations in drilling, however. Some used the reports to supplement their own wildcat hunches, but the total impact of geology on the oil industry in these states was slight.

James Donnell of the Ohio Oil Company thought little of such background work. He not only insisted that the only true discoverer of oil sands was the drill bit—he had foremen prepare the drilling logs after the drilling was done, leaving the poor driller to guess at many of his entries.

"The oil geologist must have all the knowledge and skills that an Eagle Scout should have, plus special education and training in the scientific field of geology. Star study, map making, pioneering, camping, marksmanship, cooking, first aid and woodcraft are only a few of the things he must learn …Added to all this he must have the courage and enthusiasm for his job that will carry him on to remote and dangerous corners of the globe."

Dr. Bela Hubbard
quoted in Our Oil Hunters

"Flapper geologists" pose for the camera in
Oklahoma in 1909. Courtesy, Donald Gould.

·FICKLE CALIFORNIA·

Every major company working in California employed geologists by 1910, and their work was generally recognized and respected. Only in Mexico were they more highly revered. The generally favorable attitude attracted even more geologists to the area. Between 1900 and 1911, more than 40 professional geologists and geological engineers were employed at one time or other in California. This was probably more than in any other petroleum region in the world—and more than in the remainder of the United States.

In 1911, Standard of California was fresh from a break with the Standard group. Fred H. Hillman had been sent from the Ohio Oil Company to build up the California company's production. Hillman was used to prospecting in Ohio and Illinois, not California. One of his first actions was to discontinue geological work and discharge all the geologists except Eric A. Starke, who had recommended the profitable Midway field. He discovered just how fickle California could be when he promptly drilled five failures in a row—at a cost of $500,000. He decided to set up geological shop once again, and in 1913 convinced the company to purchase the Murphy holdings in the Coyote Hills and East Whittier. It was a test of courage for the package was $5 million in extended cash payments, plus stock, royalties and a large drilling commitment. Starke's professional future was at stake, along with his other geologists'. If it did not pay off, he knew they would be out of a job. Fortunately for Starke, the investment proved very profitable, and the men retained their jobs.

California was a hotbed of geologists after the turn of the century. H.R. Johnson and Ralph Arnold pose in the field at McKittrick, September, 1908.

·SIX-GUNS AND SIESTAS·

Oil had been spouting in Mexico since 1910 when the Tampico fields had opened. As Americans and Europeans poured in, the nature of the villages began to change. When Charles Walter Hamilton arrived in Tanhuijo the summer of 1912 to work for DeGolyer, he encountered what had become typical camp life, complete with cantinas where customers stood on duck boards to escape the mud. One matron offered Hamilton her innocent 15-year-old daughter for 500 pesos—guaranteed to cook, clean, sew, and keep the bed warm as long as he wanted. Hamilton considered the offer—he was lonely and needed someone to teach him conversational Spanish at the time. "Still the thought of 'buying' a girl stuck in my craw; too much like slavery," Hamilton wrote. "Instead I compromised by giving mamma my laundry to wash and mend for which I paid somewhat more than the current rate. Apparently there were no hard feelings over the 'no sale'."

In the rough-and-tumble country there was always plenty of heat—from the climate and the men. Poker was the favorite game of the drilling crews, geologists and engineers, according to Hamilton, and the men went after it in a big way. One pay night at Potrero del Llano, in a smokey room filled with men sporting six-guns, Hamilton watched $10,000 in gold wagered on a single hand. It was not unusual, he insisted, "for a loser to pledge his horse and saddle, even his store clothes, before calling it quits." Hamilton himself was not above playing a hand or two. He got into trouble in a poker game at Los Naranjos in 1913 when he took exception to remarks that a tool dresser made about a fellow geologist seated at the table. "Heated words followed and both the tool dresser and I laid our six-guns on the table," he recalled. The situation might have gotten out of hand, but about that time, the camp superintendent to whom Hamilton reported stepped up.

He called for them to put away their guns. "You soreheads can fight it out with fists tomorrow (July 4th) in front of all the men in camp," he told them. Hamilton did not sleep well that night thinking about the tool dresser, who was large and rugged.

The next morning the superintendent called Hamilton into his office and asked him to deliver an urgent message to a camp 20 miles away. "Since it was raining I could not do any field work

anyway so I might as well spend my time in the saddle as sitting around camp," Hamilton wrote. But he felt he was obligated to finish the argument with the tool dresser. The superintendent assured him he would square the affair with the tool dresser. Sure enough, when Hamilton returned to camp, the tool dresser was friendly and apparently satisfied with the outcome.

Hamilton's first job was to map the geology and topography of Vinazco. He set out by launch, then transferred to a small four-wheel dolly car which ran on the railroad but was drawn by two mules. When the railroad ran out, he resorted to horseback. At one village, the natives were hostile until he managed to fix an old woman's handcranked sewing machine. He got little sleep, however, because the pigs kept rubbing their backs on his feet and on the underside of his canvas cot.

If a geologist was lucky enough to have a Mexican guide and assistant, he had to deal with a cultural difference that often made him pull his hair. He wanted to be up at the crack of dawn before the heat of the day set in. But the *mozo* often had a different attitude. The mozo felt God had given the night for sleeping and the day for resting. He might be awake at dawn; but he was far less likely to be ready to move.

Hamilton's crew consisted of himself and one native to hold his rod or cut trail while he made compass-and-pace traverses—except on Sundays and holidays when his *mozo* had time off. On one of these "no-helper days," he decided to scout a nearby small stream. He set out alone with compass, notebook, pistol and machete. By noon, he had worked his way up the little valley along the dry streambed except where deep pools forced him to cut through the vegetation along the banks. At one water hole, Hamilton had to take to the bank in a jimbal (large clump of giant bamboo) which stretched along both banks for some distance. The giant bamboo had needle-sharp barbs along the stem which could rip up flesh and clothing. The only way to get through was to cut and slash with a machete. "I had gotten well into the *jimbal* when suddenly I felt something soft, yet heavy, slithering down over the edge of my Stetson and onto my shoulder. I crouched, looked up and saw a large snake slipping tail first off its platform of matted bamboo leaves." Hamilton backed out until he could tell that the snake was a young boa about eight feet long. "Not caring to argue with such a formidable intruder....I dispatched him from this world with pistol and machete," Hamilton wrote. "Now and then I still dream of that boa."

·HELL IN THE GROUND·

In 1913, Hamilton was sent to Dos Bocas to observe a huge crater which covered over 40 acres and from which hot salt water poured. Everything nearby had been killed by the hydrogen sulphide gas. He was to measure the temperature patterns in the area. "The great bowl had a high side flanked by a forest of dead trees and a low swampy side," he wrote. "Through this swamp poured the overflow of hot salt water....The entire surface of the dark

Conditions could be rough even for a horse in such areas as Hill County, Montana. In 1908, the U.S.G.S. sent their "wrecking crew" to pull a horse out of the mud. The horse had been stuck on the banks of Milk River for half a day before the crew found him and dragged him out with a team. L.J. Pepperberg. U.S. Geological Survey.

On October 21, 1908, winter caught one group of geologists by surprise. They dubbed their operation "Camp Snowbound." It was located at Arbogast's ranch 15 miles northwest of Harlem in Blaine County, Montana. V. H. Barnett, U.S. Geological Survey.

fluid in the crater was in a constant motion of currents and eddies, whirlpools and blows of oily muck, hot salt water and evil smelling gas....It was an awesome sight. It smelled and looked like I imagined hell might look and smell.''

Hamilton measured the temperatures with a thermometer tied to a fishing line which he threw out, left five minutes and reeled back in. It was the only way he could get to some areas. On his last day of recording, he moved to the edge of the crater to take temperature readings below the high undercut banks. "Cautiously, I approached the edge of the crater and took a position near an old tree stump in order to have ample room for my long overhead cast," he wrote. "Twenty feet below, the oily, hot water swirled and eddied under the bank....I made a mighty heave that flung my heavy thermometers far out and down into the fluid inferno." He braced himself and held his pole as steady as possible. Suddenly, without warning, the bank edge where Hamilton was perched gave way and slithered off. "Instinctively, I reached up and groped with my hands as I slid downward. Fortunately, my hands struck a dead tree root, which the earth slide had uncovered. To this root I clung in desperation, first with one hand and then with both, my feet and body dangling over the evil hot water below." Hamilton knew that if he fell, no one would know what had happened to him. He cried out for help but there was no one to hear. "For two or three minutes I dangled over the cauldron before I could marshal enough strength to begin to slowly pull myself upward," he wrote. He worked his way up, hand over hand, all the while fearing that the root would break. "Dirt sifted into my eyes, sweat poured from my body, my head was splitting from the sulphur gas fumes and all my muscles ached from my dead weight and the tension of the situation. At long last I lay sprawled on the ground." He inched his way slowly from the crater's edge to firm ground and lay face down, trembling all over. "Never in all my life," he wrote, "have I been so completely terrorized."

Hamilton was lucky. Many of the wells in the Mexican fields produced dangerous gases which could knock a man out cold or cause a painful inflammation in the eyelids. Although the companies kept doctors in the field, they found that the eye irritation was solved by a simple homespun remedy. A slice of raw potato held against the shut eye proved more effective than medicines or chemical solutions.

Despite their willingness to undergo dangerous conditions and their fancy terminology, not everyone was convinced that geologists were worth their salt. "All your geologists' locations have so far resulted in dry holes," Sam Weaver reputedly remarked, "and I've listened to the BS of geologists so many years that I thought we'd try heifer dung for a change (to locate a well) just for luck." Whether the heifer dung, Weaver's eye for the country or his proximity to Potrero (only twelve miles south) was the ruling factor, he became the richest oil man in the Alamo field.

"The gang I spent the past summer with in eastern Colorado, near the Kansas line," one young man wrote to Miss Mary Hamman, Kingman, Indiana, on February 2, 1910, from Denver, Colorado. "Hope to be a proffessional some day. A.J.H." Blakey Group Collection.

Transportation was still primitive in 1911 when surveyors broke camp near Old Daviston, Perkins County, South Dakota. The first World War came and went before most geologists were lucky enough to switch from foot and horse to gasoline-driven automobiles. D.E. Winchester, U.S. Geological Survey.

**Geologists pack for a day's work near
Twenty Mile Creek, Niobrara County,
Wyoming, in 1910. D.E. Winchester, U.S.
Geological Survey.**

The University of Missouri-Columbia's
summer field-camp party breaks camp on
the Big Popo Agie River, Wyoming, in
1911. Tom Freeman, University of
Missouri-Columbia.

A typical U.S.G.S. field party poses around a plane table in South Dakota in June, 1911.   D.E. Winchester, U.S. Geological Survey.

Amenities of civilization were often few and far between. Surveyors in South Dakota in 1911 helped each other out in camp when possible. D.E. Winchester, U.S. Geological Survey.

A gas gusher spews in Matagorda County, Texas, in the Big Hill oil fields, "reached quickly and comfortably via the Santa Fe and Cane Belt railroad."

(Opposite)
North Dakota was still the Wild West when the U.S. Geological Survey conducted the Marmarth Lignite Survey in 1911. Left to right: A.C. Collins, (?) Hunter, W.A. Price, C.J. Hares, T.A. Birch.

This field off the south shore of Grand Lake, St. Marys, Celina, Ohio, may have been one of the first offshore oil fields. The photo, taken about 1915, shows the wells interconnected by boardwalks. James A. Noel/AAPG.

Conditions in the camps of British
Columbia were not much different than
those in the U.S. boomtowns. Town of
Tete (Jaune Cache?) about 1913.
American Heritage Center, University of
Wyoming.

Mexico had been politically unsettled for years, and it grew more so as bandits and political leaders proliferated. Geologists were often caught in the turmoil simply because they were in the wrong spot at the right time. In February 1913, Charles Walter Hamilton and his assistant, Burton Hartley, put up in Chontla while surveying from northern Veracruz to the Isthmus of Tehuantepec. They were almost killed by the half-breeds of the town when news came that President Francisco Madero and Vice President Jose Maria Suarez had been assassinated in the streets of Mexico City and that the American army intended to invade Mexico.

Victoriano Huerta succeeded Madero, but he was not popular with the people and by fall, four guerrilla leaders had taken to the fields—Venustiano Carranza of Coahuila, who controlled northeast Mexico; Pancho Villa in Chihuahua and Durango (northwest Mexico); Emiliano Zapata in Morelos and Guerrero; and Alvaro Obregon of Sonora who controlled the fiercest fighters in Mexico—the Yaqui Indians. They were loosely united against a common cause—their enemy, Huerta; and they were none too friendly to foreigners whom they felt might be taking over Mexico. The Zapatistas were really peasants who worked in the fields and took time off occasionally to conduct a raid. The others were more military-oriented. Boys under 10 years of age served as buglers and sentries. Those over 10 took their place in battle. A woman who accompanied her man often replaced him when he was shot, and many a troop had a famous lady captain. Carranza initiated his Veracruz campaign with a small force under Candido Aguilar, who moved against Tuxpam northward to occupy the South Country oil fields.

"I wonder if you all have been getting the latest war news in the State papers," Hamilton wrote his parents November 23, 1913. "If not, probably you are much more at ease mentally than otherwise. It seems that even the Mexican newspapers are ignoring the recent invasion of the Rebel forces into the Oil Fields of Veracruz. Their ignorance is due, no doubt, to the fact that if the real condition of affairs became public property, invasion would result. To be brief, the last ten days' situation has been as follows:

"The 'Carransistas' under Gen. Candido Aguilar moved upon the Port of Tuxpam where they engaged the Federal troops. However, owing to the Rebels' lack of any artillery, they were repulsed, even after gaining parts of the town, and fell back onto our oil camps, making Tanhuijo their headquarters. During the Tuxpam fight, it was reported that the Mexicans had hired on a tugboat flying the British flag and carrying an English Vice-Consul on board—this news was wired to the U.S. Battleship *Louisiana*, which, on hearing same, immediately embarked 70 Marines with machine guns on lighters, and started to land troops, but, before the boys could land the real truth was flashed to the ship and the sailors were accordingly wigwagged to return. If the Marines had landed, old Uncle Sam would have had to invade at once. Aguilar, after establishing his 500 or 600 men in Tanjuigo, sallied forth and proceeded to take Tamiahua and every oil camp in the vicinity, Los Naranjos included. This done, he took over our telephone line, and then tied up every launch the company had. When the

"unless men can believe that there is more oil to be discovered, they will not drill for oil... Where oil is first found, in the final analysis, is in the minds of men."

Wallace Pratt

Distilling outfit used to test oil shales in the field; taken in the De Beque special quadrangle, Mesa County, Colorado, in 1913. E.G. Woodruff, U.S. Geological Survey.

Rebels had all the gringos and the oil properties at their mercy they very politely requested horses, arms, etc. and $100,000.00 each from the Aguila and Huasteca Petroleum Companies—all of which was given them. So that is the way the affair now stands—the Rebels are still in all of the oil camps, literally living off the oil companies.

"All the time these events were going on, I was in the field trying to work—every day I met Rebels and exchanged courtesies with them, while, at night, I slept well after having my man hide my horses and saddles in the monte. However, one day they did take my little pacing horse, but I bought him back for ten pesos. As I said above I was trying to work yet, in fact, I did not do much. I became acquainted with General (Doctor) R. Cardenas and easy like, requested a written pass from him in order that I might not be molested—he refused—and the next day I was stopped in my survey. So, we soon pulled up stakes and beat it to Tampico.

"Of one thing we are sure, so long as the States do not intervene there is little or no danger to foreign life—but should Wilson finally decide upon such an action, me for a gunboat. Tomorrow I expect to go back down into the country with Mr. Dewey (General Supt. of the camps) in his fast boat, the "Peggy." We intend to ride thru the camps and make a decision as to whether or not it is safe to go on with field operations at present."

Distilling outfits such as this one were used to test Green River shales in the field. D.E. Winchester, U.S. Geological Survey.

·INNOCENTS IN THE JUNGLE·

Americans had been searching the South American jungles since the early 1870s although it was usually limited to the hardier variety of geologist. Harold Beach Goodrich was sent by British interests to investigate oil possibilities in the upper Amazon River of South America around 1910-11. "The high Andes were crossed in snow and ice," he wrote many years later, "rapids were shot, and broad rivers traveled by raft and canoe, tropical jungles traversed on foot or on mule-back, the calls of parrots and monkeys, the flies and mosquitos, and oh! the jungle fever!" Which of the rigorous events actually broke his health, he never knew. But his health was undermined by the treacherous travel.

About 1911, Ralph Arnold directed the economic geologic survey of oil resources of northern Venezuela and Trinidad. At that time, Venezuela had one oil well producing a paltry 40 gallons a day. The survey, which was made for General Asphalt Co. and its subsidiaries, lasted five years and was the most extensive of its kind ever undertaken in South America—and perhaps the world. It employed 52 American geologists during the work, not including numerous engineers, drillers and other support personnel.

In 1913, Kessack Duke White explored Venezuela for the Barber Asphalt Co. At that time, as he put it, "political parties argued their differences in a burst of gunfire (and) everybody dived for the doorways."

White's first expedition from Maracaibo to Machiques consisted of four or five burros, two "nigroid peones" and some cases of canned food. "We were all on foot as we were mapping by Brunton Compass and pacing," he wrote. "It was considered that pacing on horseback was much less accurate." They failed to reach a waterhole the first night so they slept under the stars. Before noon, the next day, they reached Pozo Verde (green hole or well), a small pond filled with green algae.

"We rested a couple of hours and decided to push on much to the consternation of our two peones. The toothless one said "No ah Awah." We thumbed our little dictionaries—neither of us could speak a word of Spanish—without results or understanding. Finally one brought a rusty can filled with the green slime, shook his finger over it and said No! No! No! We concluded there was no water ahead that could be reached that day. We did not believe them, but pitched camp by the green hole. We left next morning at day break, marched all day and just before dark came to a hog wallow. We kicked out the pigs, scooped up a bucket of very muddy water, strained it through a towel, boiled it, added tea leaves and drank it.

Kirtley F. Matther on his favorite mode of transportation while working for the U.S. Geological Survey in the San Juan Mountains, Colorado, summer, 1912. Note the chaps for riding through brush, his typical bedroll, and rifle for hunting. All in all, he is traveling light.

"The next day, we were traveling the camino real (sic), there was a loud growl from the jungle. (Leonard G.?) Donnelly, the other geologist of the party, wanted to investigate, I thought the main highway bad enough. He insisted he was going, I grumbled and complained, but went along. We climbed through the barb wire fence bordering the trail, grasping our little 32 cal. Colt pistol. As we went into the jungle the sounds grew louder. While we walked Donnelly continued to suggest various animals including an alligator. Then, we looked up and saw the source—Howler monkeys. Disgusted, we returned to the Camino Real covered with ticks.

"We arrived at Machiques, the capital of Perija. We had orders to go to the Rio Perija. This was dangerous Indian country. The governor told us if we went beyond the Rio Negro he would send soldiers to place us under arrest and would return us under arrest to Maricaibo as incompetent to be trusted alone. He also said he did not want to have to report our deaths."

While White was working for Barber Asphalt, Ralph Arnold and his wife made an inspection trip. "They arrived in (Maracaibo) where his boys in the field were assembled," White wrote. "The Company had a fair size yacht on Lake Maracaibo and an evening on the lake was planned. The Arnolds did not drink so each of us unknown to the other arrived with a bottle of grape juice under one arm and a box of macarrons (sic) under the other.

"So we put off with a diet of grape juice and macarrons; enthusiasm died, the party became sticky so the Captain cooked up a story, since...the night was black with no moon the danger of collison with a fishing boat was too great. So we put back to shore. We said politely good night, thanks for the evening and trip and as soon as the Arnolds were out of sight headed for the oasis on the Plaza.

"Well it started with beer and the founding of the Society of Intrepid Explorers. Then someone had ordered a bottle of champagne which was on a rickety table. The table upset, the bottle smashed on the floor, who caused the accident could not be determined so each one had to buy a bottle. The former society was considered too plebian so the 'Christ Like Society, was christened. It was a wet night."

·MEN WITH EAGLE EYES·

Many self-styled experts—including the geologists of the day—were positive "not a barrel of oil" would be found near Cushing, Oklahoma. The town was a sleepy little wayside spot without phones, electric lights or cars. Scouting in the countryside was done by horse and buggy, and any scout worth his salt stuffed enough grub under the buggy seat for a three-day tour. A penciled map scrawled on wrapping paper usually completed the average gear. When oil flowed over the derricks in 1912, Shamrock and Drumright sprouted up in the field. Cushing, 10 miles away, had the railroad, however, and the oil business ran over the town sidewalks—storage yards for pipe and machinery, lumber yards for rig-building, restaurants and hotels, jewelry shops and auto dealers. But there was no sewer system, food was in short supply and housing was impossible. Typhoid was common as well as meningitis. There were, as one man wrote, "vultures, harpies and the riffraff of the country." It was not uncommon for scouts and wildcatters to go into nearby towns and lease up every available horse and buggy and close the stables with a padlock. This kept anyone else from getting around, investigating possible drilling or lease sites. "Those were the days when the best and the worst in men were brought out," wrote James McIntyre. "Men lost all sense of proportion and perspective. Things were said and done in heat and hate, which some of them would like now to forget."

Until that time, most of the successful oil exploration had been conducted by corporations with large resources. Cushing proved that the little man could also strike it rich. The Exchange National Bank in Tulsa—organized as a venture-capital bank for oil deals—provided the backing for many of them. The field itself yielded so many fantastic gushers that Oklahomans claimed it made a crop of 100 millionaires.

Even Oklahoma could be cold, and a fire was welcomed on the first student topographic party in the Arbuckles. Longjohns were aired on the top of the tent.   Western History Collections, University of Oklahoma.

Students prepare a topographic map in the Arbuckle Mountains. They are using the standard Alidade and plane table; an associate is probably holding a calibrated rod on a nearby hillside, and the surveyor is "shooting" him through the telescopic Alidade. Western History Collections, University of Oklahoma.

steamships and railways converted from coal to crude. By January 1913, oil was up to 83 cents a barrel—and the Cushing Field was spouting 20,000 barrels a day. When they drilled deeper and tapped the Bartlesville sand, it jumped to 60,000 barrels a day.

As the value of oil increased, so did the value of the geologist—at least he had a chance at getting a job. In June 1913, Charles Gould advised young Everett Carpenter that the Quapaw Gas Company of Bartlesville was hunting a geologist. Carpenter wasted no time in making the application. "The Quapaw Gas Company was not among the most famous names in oil and gas of the nation," Carpenter wrote, "but we reasoned that if it had the word "Company" following its name it must be rolling in wealth....It was then that it became apparent that all corporations were not rolling in wealth. I was given an office over a grocery store, equipped with a kitchen table and chair, and told that any additional furniture would have to come later. I had none of the tools so necessary to a geologist such as a compass, aneroid, hand level, etc., but I did soon get a compass and hand level. For transportation, I had my own two legs. Automobiles were just then coming into use but only the officials had them. The roads, when there were any, were single track lanes built for the use of horse-drawn vehicles. The easiest and surest means of travel was by train. There were livery stables in all towns where horses and buggies could be rented. All towns boasted of a hotel, but usually the rooms were already occupied by many small nonpaying, but permanent guests. Geologists of those days were a hardy lot and it

The telescopic Alidade, plane table (see p. 97) and stadia rod (see p. 154), used by 2-man parties for mapping structure, were responsible for finding many oil fields--especially in the Mid-Continent. Exxon Corporation.

The Brunton compass--an inseparable companion of all field geologists. In addition to its normal function as a compass, it is used for sighting, signalling, reading strata dips, as a hand level, and other functions. Exxon Corporation.

took much to discourage them. The usual procedure was to go to the nearest town and start walking. If it was inconvenient to get back to town it was not uncommon to put up for the night at a farm or ranch house.''

That same year, Gypsy Oil Company organized a geology department under M.J. Munn, and E.G. Woodruff was employed by the Texas Company. Carpenter began work in the Augusta, Kansas, area, mapping the Fort Riley and Winfield limestones in August 1913. They were "mapped by pacing the distances from numerous points on the outcrops to Section lines and corners....It was a noble map—one that would warm the heart of the most cold blooded. There was just one thing wrong with it—the general manager could not understand what was meant to be shown....Still a good enough picture was made to enable me to convince him that a detailed survey with a telescopic alidade was necessary. But as it was still forbidden to spend any money even for an alidade, I did not get permission to buy one.

"It took considerable scheming and a lot of nerve to get around that hurdle, but I finally succeeded. I waited until all the officials were to be away from the office for two weeks and placed an order through the purchasing agent. When the officials returned the alidade had already been delivered. Of course, I was called on the carpet for a rather severe lecture but no economic sanctions were imposed on me." J. Russell Crabtree was employed to help make the survey. When T.N. Barnsdall and Alfred J. Diescher looked at the maps and listened to Crabtree and Carpenter, Barnsdall asked if they thought it was a good place to drill because the contour lines ran around in complete circles. "When assured that this was the belief, he gave his consent to whatever it was that was desired. This was the first deal that I know of that was made purely on geologic evidence."

Because of the field's success, the stock of geologists zoomed upward, according to Carpenter. After Augusta was brought in, Carpenter reported, we were "quite generously called 'the men with the eagle eyes.' "

By 1914, Gypsy's petroleum geologists numbered 30 men. Their work in the field made some competitors realize the advantages of petroleum geology, but it was not a widespread consensus. C.L. Severy was hired as Carter Oil Company's first geologist. "I don't think you will do us any good and I will damn well see that you don't do us any harm," J. Edgar Pew told him. Pew meant it, too. "He would not buy me an aneroid barometer," Severy recalled. "So Dorsey Hager, one of the first consulting geologists in Tulsa, loaned me one until I could find a good one to buy for myself. Soon after that I needed a plane table and alidade. I applied for it but did not get it until I persuaded young Bedford, the treasurer of Carter who was in Tulsa on an inspection trip, to buy me one. It wasn't that they were stingy or did not have the money. It was simply that they could not see spending good money on a crazy idea of a geologist."

C. Max Bauer, a "grand geologist," according to Frank Gouin, taught Gouin how to "gun the sun." Bauer poses in the Tonkawa field, Noble and Kay counties, Oklahoma. At the time, they could not get enough evidence from surface mapping to prove they were on an important anticline. It was not until later core drilling by Marland Oil that the presence of the Tonkawa anticline was confirmed. AAPG.

The Midwest oil fields were booming. Geologists armed with map and planetable combed southern Oklahoma as the U.S.G.S. began subsurface mapping in the Healdton field in 1914. Kenneth C. Heald was assigned to the area. "I hired a buckboard with a team of horses and a young chap who was recommended to me by a local minister in Duncan, and we started to map the Healdton field," he wrote. "I quickly found out that there was very little in the surface geology that would help me depict the structure of the field and I remembered that M.J. Munn...had depicted the structure as revealed by the elevations on the top of the oil-bearing member....I, therefore, decided to apply this to the Healdton field and meticulously determined the sea-level elevation of the well collar on each well." Heald obtained drill records of 270 wells and prepared a contour map and a stereogram (the first of its kind in a U.S.G.S. publication) which showed the structure on one of the oil horizons by 10-foot contours.

Up in Kansas, geologists were trying to decide whether or not the area had much oil. Back in 1909, state geologist Erasmus "Daddy" Haworth had published volume nine of the University of Kansas Geological Survey, which discussed oil and gas resources of Kansas. "We set it down in Volume IX that 'where valuable production has been obtained there are practically no seeps of either oil or gas,' " explained Wallace Pratt, who graduated from KU that year. " 'If oil and gas is escaping at the surface, it is a sure indication they do not exist in the depths below....in very large quantities.' " Another gospel truth by which Kansas geologists swore was that the Mississippi Lime was the absolute floor in Kansas. "Even to have suggested the possibility of encountering oil in a lower horizon, the now copiously productive Ordovician, for example, would have stamped one as a crackpot."

Around 1914, a wild-cat well reported granite "at a depth of 1100 feet, or so, right in the middle of the state," Pratt wrote. "We were not only skeptical, we were indignant. We denied (it); and when the driller, under our own supervision, bailed out of the well fragments of beautiful pink granite, we charged that he had planted the granite in there himself! So little were we prepared for anything like the now classic buried Nemaha Mountains, traversing the width of our state just under its surface: still less could we have conceived that such a buried mountain chain, under the sediments spread over it by encroaching seas, might flank itself with porous reservoirs in which important pools of oil might be trapped.

"A little later I tried to persuade the chief geologist of a large oil-producing company that he should search for oil in Kansas. I got nowhere. He stated flatly that Kansas was composed largely of limestone and shale whereas oil was usually found in sandstones. Accordingly he sought a stratigraphic sequence of sandstone and shale in preference to limestone and shale and on this principle he declined to hunt for oil in Kansas."

Frank Gouin poses for the camera--"a silly geologist trying to act like Balboa discovering the Pacific," he remarked. AAPG.

Not only were there new techniques, but there were stories of new kinds of equipment that could help identify favorable conditions for oil. Everette DeGolyer had opened his own consulting business in 1914 and promptly ordered a new model of a torsion balance then being used in Hungary by Roland Eotvos. DeGolyer would wait many years—and the petroleum geology industry be put on hold—while men fought a war in Europe and prevented delivery of the new equipment.

Even without such equipment, geologists were doing their work better than ever. Oil was everywhere in 1914. In Canada, everyone had Turney Valley oil fever. One Calgary business syndicate was studying far north of Alberta in an area where J.K. Cornwall of the Northern Trading Company had noted oil floating along the banks of the Mackenzie River below Fort Norman. About this time, Dr. T.O. Bosworth, an English geologist, just "happened to be in Calgary on his way back to England from South America where he had been employed by the Shell Oil Company." Bosworth examined the properties and reported that the seepages were not that impressive, but certainly the Fort Creek shales and Beavertail limestone were remarkable. Norman Wells, however, was not a particularly inviting country. In summer, it was said that mosquitoes were big enough to shoot down with a rifle. In winter, temperatures sank to 60 below.

That same year, Imperial Oil Limited, a subsidiary of Rockefeller's Standard Oil, established an exploration and production department and set out exploring—from the sub-Arctic of northern Canada to the tropics of Peru. Bosworth was hired as chief geologist. He purchased the Norman Wells properties from the Calgary syndicate. But they remained dormant, waiting for the right man for another five years.

The overabundance of black gold had a dire effect upon the industry. "Glooms have got the oil business," the *California Oil World* reported in September 12, 1914. "Talk to an oil man these hard days and he sees nothing for the future. There is no market, no possible expansion of the market, no chance for improvement, no visible ray of light."

Job hunting was tough on the West Coast. When Walter English graduated from the University of California at Berkeley, he took it upon himself to visit W.W. Orcutt, vice president in charge of exploration for Union Oil Company of California. Orcutt was so well known for finding oil that it was sometimes said that when he went to build a house, he had difficulty finding a piece of land without oil under it. "I'm sorry I can't give you a job," Orcutt replied. "Union Oil already has a geologist—as a matter of fact, I am a geologist myself, so we are completely supplied with geologists."

Frank Rehm was luckier. He was a big, good-natured, companionable man who churned his way through life with easy unconcern and evident enjoyment. He began his studies at Stanford University in early 1912 and should have graduated in the fall of 1915. Just prior to the end of his final semester, however, he went on a field trip to complete a geological paper that would have earned him his degree. Fate intervened, and he met a young lady

Oil shale was already seen as a possible energy resource back in 1915. U.S.G.S. men sample a bed of oil shale near Green River in the Table Mountain quadrangle, Sweetwater County, Wyoming. D.E. Winchester, U.S. Geological Survey.

with whom he fell madly in love. In the tempest of infatuation that followed, he was not much interested in geology. In due course, Frank and the young lady were married.

After the wedding was over and he settled back down to earth, he began searching for his geological notes. But the young lady was an orderly creature and when she took up her domestic chores, she promptly cleaned the Rehm household from attic to cellar. The first thing she tossed was his rather messy looking sheaf of geological notes. It might have been disastrous for someone else, but Rehm was hired by Shell Oil without the official sheepskin. (He would not get around to completing his degree for another 27 years.)

Shell was hiring because they had begun major geological investigations. One of the projects was to produce a detailed geological survey of the Ventura area. Every dip taken in the field was shot in with a transit and rod. In that way, they obtained both the exact location and elevation. The resulting final subsurface topographic map was projected to a depth of 5,000 feet—greater than any producing oil measures then known. The survey was the first of its kind in the state and proved startlingly accurate over the years.

·WILD WEST WYOMING·

"In 1914 I was working with Charlie Orchard hunting for new oil fields in Wyoming," wrote Ralph Arnold. "Charlie was very familiar with the country, having been over it in connection with his cattle business. He took me to Grass Creek. I examined it and we leased all the Government land that was open. I went back with him and made the location for the test well....We put down the discovery well, which came in a good producer. As soon as oil was struck the Midwest (Refining Company) officers came in a private car to Worland. They employed a gang of gun men to run us off our leases and we came back at them with another gang of gun men to protect our leases. It looked as if we were going to have some real wild west shooting.

"I learned that Cass Fisher, my former partner, was with the Midwest 'brass' and I hurried down to their private car at Worland and we worked out a deal dividing the leases. War was averted."

In 1915, the U.S.G.S. issued Bulletin 581 which included a paper on the Big Muddy dome, Converse and Natrona counties. "I was one of the first to secure a copy of this bulletin and to realize the importance of (V.H.) Barnett's conclusions," said Ralph Arnold. "I set about at once to secure capital with which to lease all of the State lands on the structure and to acquire as many government leases as possible. I induced my friends F. Julius Fohs and James H. Gardner, geologists of Tulsa, Okla., to check my location for a test well and to secure money with which to drill a well. They interested Humphreys and Whiteside of Duluth, Minnesota, in the project. A well was drilled and brot (sic) in the field. It took over a year to finish the test-well, so that our leases ran out on the State

U.S.G.S. men sample oil shale near Green River in the Twin Buttes quadrangle of Sweetwater County, Wyoming, in 1915. D.E. Winchester, U.S. Geological Survey.

lands, and state officials and others on the 'inside' were able to take over the State lands and leave me and my associates out of the picture. I give Barnett credit for finding the field and Fohs and Gardner credit for locating the discovery well. Humphreys and Whiteside drilled the discovery well. I will take out 'two bits' worth of academic credit for calling the attention of the oil industry to the area.''

Fields were heating up in Kansas and Oklahoma. When J. Russell Crabtree made a detailed survey of the Eldorado area east of Wichita and published it in cooperation with the U.S. Geological Survey and the Kansas Geological Survey, it was dubbed the "first wide area geological survey ever to be made." The Eldorado Field was opened up in 1915. It was the first large field discovered by petroleum geologists using proper geological techniques. It was big enough and profitable enough that Empire Gas and Fuel executives took a second look at the idea of petroleum geology.

Another promising area was northeast Oklahoma—Osage County. In the summer and fall of 1915, Kenneth Heald worked the area alone. Since he did not have a helper, "I did what many lone geologists had done before and have done since," he wrote. He lashed his stadia rod to the side of his car and circled it, "taking points by setting up the plane table on or near outcrops." (Later, when Heald married, he taught his wife to run the alidade.)

The University of Missouri-Columbia, summer field-camp party, explored the high Wind River Mountains of Wyoming in 1916. Left to right--McFarland, Groves Gillson, Schnapp, a Mr. Robins, and Professor E. B. Branson. Tom Freeman, University of Missouri-Columbia.

The summer field-camp party of
University of Missouri-Columbia explored
near Fremont Peak, Wyoming, in 1916.
Tom Freeman, University of
Missouri-Columbia.

Survey men drill oil shale near Watson for a sample to distill. Dragon quadrangle, Uintah County, Utah, 1915. D.E. Winchester, U.S. Geological Survey.

"The geologist is the keystone of a great arch involving not only his own but many other industries. No other group has done more to raise living standards worldwide. I like to quote this little statistic to my lawyer friends. Less than 10,000 active petroleum geologists, and lesser numbers in the past, have pointed the way to trillions of dollars worth of energy for the benefit of mankind. By comparison, there are millions of lawyers in the United States and hundreds of thousands in government. Their numbers are increasing pari passu with the trillions of dollars of our national, state, municipal, and private debt, plus the scores of billions of annual government waste, for all of which they are chiefly responsible. But for the grace of God, I might have been one of them."

Lewis Weeks

Distilling outfits such as this one were used to test Green River shales in the field. The picture was taken in 1914 in Rio Blanco County, Colorado. D.E. Winchester, U.S. Geological Survey.

·TWO DOLLARS A DAY—NO PAJAMAS·

By 1915, other companies were hiring fledgling geologists. When Jesse Vernon entered the University of Nebraska in 1915, he had $5 in his pocket and one suit of clothes. Short of money, he took work as an instrument man on an Empire Oil and Gas Company surface geologic party, to stretch him through. Like so many others, he became addicted to geology, and his life was forever altered.

Many of the young graduates sought employment with the U.S.G.S. or with the state surveys. If it was exciting to the young men, it was rarely lucrative. "After I completed certain courses in geology and related subjects," wrote Hugh Miser, "a covered wagon transported me to my first job. I was loaded into the wagon, together with camp equipment, for the week-long journey. For my first eight days' work, I received $8 from the state of Arkansas; the notary charged half a day's pay to certify the voucher. Immediately after that eight-day period my party chief placed me on the U.S. Geological Survey's payroll at $2 per day. That...proved to be my biggest percentage increase in salary."

E.W. McCrary was working numerous young men in the Oklahoma area for the Tidewater Oil Company. Under his supervision, field work was devoid of frills and luxuries. It was said that his men absorbed most of their geology through the soles of their boots. J. Elmer Thomas, who had been accustomed to the amenities of Chicago, went to work for him in 1915, according to Grover C. Potter. Thomas's service ended abruptly, however, when he asked McCrary how many pair of pajamas he should take along on a field assignment. McCrary replaced him with a boy from Arkansas who did not need pajamas or other such excess baggage.

·TURMOIL IN TURKEY·

On June 24, 1914, Kessack Duke White—ever the intrepid explorer—sailed from New York on the steamer *La France*. He was in the hire of Standard Oil of New York, and he was headed for Turkey.

"Heavy fog in harbor passed *Lusitania* that was hung up in fog on way out," he wrote in his diary. "Crowd mostly foreign met no one outside of party. Trip is very calm," he wrote the next day. "No one sea sick, tried a tango tea in afternoon and dance at night no one very enthusiastic." White made friends with one family, the Farons (or Faraons) in whose train compartment he rode to Paris. "Along the route children ran after the American train and we threw them pennies," he wrote. One of the traditional problems was baggage. "We had 22 pieces or about a ton and a half," he wrote, and we had "considerable trouble getting it." While in Paris, White attended the theater where he saw "Mistenguey and her dancing pardner Maurice Chevelier."

July 4, White set out for Constantinople aboard the Orient Express. "(It) is not so bad but the trains here are a joke compared with those at home," he wrote. "The great difference noticed is the intensive cultivation in Germany, France, Austria, the country being a continuous park, compared to U.S.A.

"From Serbia on, things were primitive and the houses in the country were mostly huts of mud and sticks with tile roof....Serbia, in those days, was rather wild. Suddenly the mustard jar on the table exploded in the dining car while we were seated. Some one had taken a shot at the train from the hills."

In Constantinople, he managed a little sightseeing while he "did a little packing and a lot of buying....In the afternoon went to Therapia the summer colony for Constantinople. The Bosphorus is beautiful with its swift current and steep sides. The company launch is a dandy and the trip thoroughly enjoyable....Saturday pack equiptment (sic) all day." There was a hint of disappointment when he noted that "Mr. Faraon arrived in town but his daughter was not with him."

A geologist takes oil from a Nitig oil seepage in the province of of Erzeroum, Turkey. Kessack Duke White and Raymond S. Blatchley worked the area in fall, 1914 for the Standard Oil Company of New York.   American Heritage Center, University of Wyoming.

From Constantinople, White sailed to Trebizon (Trebizonde) where they met with the Governor General. "The hall was in bad need of repair," he wrote. "A number of men and women were standing around, I suppose to get their troubles straightened. We had no trouble to obtain an audience. Presented our letter from the biggest man in the country, the Sultan is merely a figure head, while the Minister of Foreign Affairs runs the government. The Governor General immediately unbent and gave us what we wanted, Gendarmes. The country was originally thick with bandits, but they have been cleaned out. In the evening we went to a moving picture, which was very good, Pathe, guests of the owners, and had ices afterwards."

From Constantinople, White, Blatch and Kobbe set out for Erzurum. With them were several wagons pulled by three or four horses abreast. "Blatch was sore from the horse," White wrote. "We stopped at 8 o'clock for lunch and rest horses. While at the village some one stole my pearl handled knife. One of our wagon boys was accused, and probably did it fooled around about an hour. The driver would have beat him to death, if Ortin had not stopped him. We left the boy as well as the knife and proceeded on. Crossed the mountain pass 6000' above sea level." When the group were down the other side, they met up with the wagons. "To our surprise the boy we left was driving one of the wagons....It seems he had followed, caught up one of the drivers got mad and cut him in 3 places, he complained to the Gendarme who held up the wagons for explanation so to get in by dark, they let him come. He was sure persistant, and I believe he took the knife....We agreed to allow him to go to Biburt but no farther." They stopped at the hotel at Biburt "which was big enough but very dirty and the beds fairly alive with bugs and fleas, I got thru the night some how but was badly bitten in the morning."

The trip that had been estimated for a week took over three, and they finally arrived in Erzurum July 26. The group quickly organized and began making geological trips. On August 5, White developed a problem that frequently plagued the world travelers. "Have had diarrhea was in bad shape yesterday," he wrote. "Staid in camp today....Did a little plane table work in the afternoon." This time, White's pistol disappeared.

Because of unstable conditions, they traveled with gendarmes as guards. "War talk is everywhere," he wrote on August 9. "At Chyms a Kurdish town every man up to 45 year has been taken....The country has no provision with which to feed an army. The crop are ready to harvest and no one to do it except women, old men and boys. Each conscript is required to take 5 days food with him. The whole country is being ruined. the crop are ruined. Everyone is talking against it but it does no good."

If there was major trouble, the expedition expected to hear from the New York office. But there was nothing, and they continued to work. Almost a week later, the gendarme and a packer arrived with mail and supplies. "Only a telegram from the office saying that everything was quiet in Constantinople, but to use our own judgment as to whether it was best to come in or not." Meanwhile, local conditions were worsening. "We are having a little trouble with the authorities, they have refused us a gendarme," White wrote, "but we wrote a letter to the governor which will probably bring the police to time."

The group camped near Hassen Kale, about 14 hours by horse from the Russian border. "This morning the head of the village...called and informed us that if our horses were stolen he would not be responsible as horses were being stolen all over the country, so our men sleep with them."

On Sunday August 16, they went out in two geological parties, "Blatch and myself each with an interpreter and guide. We continued as far the oil seepage, which of very good size and issued with sulphurated hydrogen and brine. ....On the return trip stopped (at) a small mountain lake that reminds me of the pictures of Switzerland, and had a bath in the clear cold water."

·ECLIPSE ON THE PROJECT·

On August 21, White went out to the oil seepage to map from the mountains. "While on top of the mountain the eclipse occured (sic) at 3:30 by my watch....I was sitting on some quartzite triangulating to some adjacent peaks and sketching the geology when I noticed that it was becoming dusk. I did not have any smoked glass, but could see part of it covered. The effect was peculiar sort of a greenish blue." The eclipse lasted for less than two minutes, but the night winds set in and the cool continued for the rest of the afternoon. That night, they entertained three Kurdish chiefs at supper. "It is Ramozan so that supper began promptly at 7 o'clock as this is the hour they are permitted to break the days fast of food & drink," he wrote. "They acquitted themselves very well and used knives, forks and spoons admirably. After dinner which was a little pretentious considering the lack of diversity of foodstuff in the region, they excused themselves before tea to play the gong one at a time after they had completed their obligation returned and staid until about 10 o'clock."

Still there were no messages from the home office. "We were just getting to bed when a messenger arrived that we had sent to Erzurum about 20 miles away. He had traveled day and night to make the trip. Brought a letter from (Cass?) giving the first accurate news we had of the war and its cause." The letter was 11 days old and had taken seven days to get by telegraph to Erzurum. It read, "Return to Constantinople immediately." "Every thing went into the air especially Blatch," White wrote. "He wanted to set out for Erzurum that night but we persuaded him to remain untill morning. All mail boats are reported taken off the Black Sea, so it looks as if we had a 500 mile overland trip to Angora the nearest railroad point to Constantinople."

Kobbe, the drilling or production engineer, together with the Armenian [interpreter] and two of the gendarmes, rode all night and arrived before the Ottoman bank opened its doors the next morning. "We had several thousand Turkish pounds credit which they withdrew in gold coins," White wrote. "That afternoon the moratorium was declared. From then on we were all heavy with money belts wherever we went...

"Turkish mobilization was declared the roads were crowded with recruits and occasionally German officers headed for Erzurum, general headquarters for that part of Turkey. Conditions became a bit more difficult." They passed many camels loaded with ammunition. By August 27, they had gotten as far as (Maden Han?). Their horses had given out, and they stopped for the night at an inn bordering an ancient walled Armenian city in ruins. "Recruits continued to pass but not in the numbers of yesterday," White wrote. "The action of the government is criminal, sick men and all are being taken...One of them was obviously too sick to travel. We gave him some medicine, arranged for him to be fed and to have a bed. After dinner he courteously approached, salaamed, took from his pocket a handfull of hazel nuts, carefully divided them into two parts, one part he offerred to us and salaamed. That gesture of offerring one half of all his worldly possessions was his way of saying thanks and appreciation of our attention and aid....The next morning we put him in the carriage and drove to the military post; explained his condition to the commandant who gave him his discharge and permission to return home.

Each Turkish soldier was allowed 33 piastres, about $1.25, to travel 200 miles— about 20 cents a day. The two Armenians with White were worried to death. "Turkey should be cut to pieces and it seems that the day is not far off," he wrote. By August 29, telegrams indicated that Germany was within 10 hours of Paris. On the 31st, White's party arrived at the pass. "Pass was clear met a smuggler...and bought a couple of pounds of tobacco....I am being passed off as usual for a German, (as) I have been taken continuously...and (as) I look (more) like one than any one else, I have to assume the role, if a real German came along I will sure be up against it."

They arrived at the Bristol hotel (city unidentified) at midnight, September 1. "They turned out fine to receive us and we each have a pleasant room. Awoke at 6 AM and dressed in real clothes which feel very good. The Armenian boat flying Belgian colors arrived today, so we will leave for Constantinople tomorrow." In the streets below, White noted soldiers marching to the tune of their strange music with "a number of bugles and a drum."

Although the boat left that evening, they did not reach Constantinople until October 14. Government red tape hampered travel, and the boat kept getting stopped. "Reached (Rodosto?) governor held us up for permits, and vissika (visa), went to Constantinople, Sunday Oct 18, cussed our attorney, who was supposed to have every thing necessary. He did not have the vissaka so that the governor, who was really acting for the Germans, would not permit us to work. Gomezian went back to Constantinople for the vissica, we by boat to Hora."

After some difficulty, the group made it to Constantinople. While there, the German warships *Goeben* and *Breslau* arrived. They had outrun the British Mediterranean Fleet by forced draught that burnt out all their boiler tubes. With the ships' guns trained on Constantinople, it was only a question of when Turkey would enter the war.

"There were several thousand German civilians employed by the Turkish government and Enver Bey later Enver Pasha, the War Minister, was German educated and pro-German.

"After a short stay in Constantinople we went into Thrace. Shallow wells had found some paraffin oil near the villages of Hora. The Germans were not happy for our presence, but we traveled the country and did our work under the protection of the Turkish Captain of Port stationed at the village of Hora....

"A messenger arrived from Constantinople with orders to return immediately. He found us in a Greek village at the foot of Mnt. St. Elias, the highest peak of the Dardanelles range. Turkey had declared war on the Allies, further work was impossible."

The men left Constantinople by train via Bulgaria, managed to catch a boat to Greece, another boat to Naples. From there they boarded a boat back to the safety of the States. With turmoil all around, and constant troubles with the authorities, White was still the ever-curious geologist. One of his last entries in his journal was a simple one-line sentence: "While waiting for a boat to the States in Naples we visited Vesuvius."

·ANY WAY YOU CAN·

The effects of the war were felt halfway around the globe and far from what most Americans thought of as civilization. By 1914, many Mexicans had become German sympathizers, and Mexico became a hotbed of international intrigue. In addition, criminals, radicals, adventurers, riff-raff—and later draft-dodgers—crossed the border into Mexico.

Some geologists continued to work despite the dangerous conditions. They were easy targets as they moved through the jungle loaded down with instruments, foodstuffs and canteens. Payrolls were almost impossible to get through. Only gold and silver were accepted, and over a period of time, the oil companies became inventive. The treasure was packed in kegs of red lead, in the cylinder heads of oil field pumps, bales of cotton waste, in pipes and stoves. It was carried by speedboats, in gasoline drums, concealed in the bilge water beneath floor boards, the tool boxes of cars or in tires. Bandits, too, became more inventive in finding it. Ex-Texas rangers were hired to guard the payrolls, but so many were killed or wounded in ambush that the remainder quit. The companies finally resorted to staff volunteers, using different people and different routes each time the payroll was due. It was a job few relished, and plenty of prayers were uttered from the lips of those who rarely engaged in such activity.

The canals of Tierra del Fuego were explored using a diesel schooner. American Heritage Center, University of Wyoming.

After Kessack White's escapade in Turkey, he returned to the states. But by 1915, he found himself traveling again, this time to Colombia. Trouble began just after he left Palermo. "The old man is a sort of major domo for General (Trujillo?)," he wrote January 28. "After thinking it over during the night, though he had consented last night, refused to act as a guide because the General had not given the order; but after much talking agreed to go." February 13 he wrote, "The entire time (Feb 2-Feb 13) spend in Lorica. Mowinkle arrived with Anderson. Mowinkle taken sick and is still in bed. Left by Gasoline launch saturday for Monteria...trifty (sic) town electric light ice plant bull ring & moving picture show. Good Hotel.

"Saturday night went to a picture show. Sunday wrote letters all day and went to Bull fight at night. Did no killing only stuck in the Banderillas.

"Monday got the pack train off. That night went to a dance which was a very pleasant affair. The girls were well dressed, some of the men had on dinner coats and so appear(ed) rather warm. Managed to get along without difficulty. Tuesday morning left

about 9:00 o'clock for the woods." White's party consisted of three pack animals, two mozos, Franke, the general and himself.

They visited San Diego, a crater-pocked area about 1/4 mile square, withover 50 vents. "The largest is a crater of plastic mud 60' across," he wrote.

March 6, White and Morrison struck out "to see the getting out of a 60" log. Tackle broke and (lodged) in the place we had been standing, rather a close shave." They caught up with the pack train and made Cordobela in the afternoon. "The playa are more than bad, the sand fleas bite like furry," he wrote.

"March 15. Monday. Rode to Puya to find out about guide, who turned out to be a crook and wanted about our entire camp, and a horse to ride to show us the mine. Returned to Murindo, went to Casarrubia expecting to give up the examination, General was there had a guide, returned to Murindo on a fresh animal; rode about 35 miles. Guide is to arrive in the morning....

"(March 16) On returning in the afternoon crossed a very bad "barra," mule became mired to its belly, so I rolled off as I could not disentangle myself sufficiently to jump,

(Opposite)
A company camp at San Sebastian, Colombia, South America, in 1915. It consisted of an office, cook house and bunk house, plus a water tank.   American Heritage Collection, University of Wyoming.

(Right)
A camp house at Aguila, a "characteristic house of the country," White noted. American Heritage Collection, University of Wyoming.

fortunately mud was covered with leaves so I did not get muddy. Landed at the bottom of a...small knoll. Guide lagged behind mule took wrong trail which fortunately landed us in camp a half an hour earlier.''

The General arrived from Murindo, but the pack train's animals kept wandering off. While natives worried with details, White pounded fossils out of the sandstone beds.

''(March 20) This morning 5 wild hogs stood and looked at me for several minutes when I came across them in the trail. This afternoon pass(ed) a tapir (burro danta) feeding in the jungle.''

By May, White had made it to Turbaco where he picked up riding animals, supplies and headed out for another excursion. On May 8, White wrote his mother, ''As you notice my

Drillers at San Sebastian, Colombia, South America, in 1915, had to overcome such natural vermin as gnats, mosquitoes, snakes and wild pigs. American Heritage Center, University of Wyoming.

Drillers at San Sebastian — Columbia S.a. 1915

letter is dated in camp, but it is probably not the camp of your imagination, for I am really as comfortable as if I was in civilization, and much more comfortable than in Cartagena....A couple of years ago the Company whose concessions we bought, spent $100,000 at this locality in an illadvised attempt to secure petroleum production to supply their refinery in Cartagena, the present camp is the monument of their failure. Every thing is still in good condition. There is a derrick, an electric light plant, boiler plant, and two bungaloos. I am living in the larger, it has an office, dining room, two bed-rooms, shower bath, kitchen and store-room, and at present I am writing on the back porch in my pajamas as this is the coolest spot. In fact it is a very pleasant life but (trouble?) is it won't last very long....

The geologist traveled as did the natives of Colombia--by canoe. A tree across the Chachira River detained the group in October, 1916.   American Heritage Center, University of Wyoming.

"It is now about five o'clock and quite pleasant I came in about three from a hot rather discouraging day as I did not find very much, rubbed down with bay-rum, one cannot wash after being out in the sun all day as it is apt to bring on the fever, had my dinner and cigar, read a story in a magizine (sic) and am now writing to you. The story in the magizine realy suggested the letter, for I had realy not intended to write until tomorrow as that was the last day of grace to catch the weekly mail. As it is now you see I may write every week. The story itself was nothing, but only how home-sick the wife of a French-man was who had moved from France to New York, but it suggested how very provincial a large number of people are, or to put it another way, how hard it is for some to become accustomed to other conditions than those to which they have been reared. Those who have lived in the States all their lives think any place is exile, those in Europe the same, and the same statement is true of those who have been reared here as unbelievable as it may seem....I was talking the other day to

the daughter of the lady who runs the hotel, who incidently (sic) is very well fixed, and is Irish, owns a couple of breweries, an ice plant, bottling works and sundary (sic) other ventures, the girl had been in school in England and lived some time in London and whether you believe me or not she prefers Cartagena to any place. Another case, the chap is President of the largest bank here, has spent considerable time in the States and Europe, he said, "If I could have what I wanted, I would have two homes, one in Nice France, and the other here in Cartagena, and spend six months in each, so you see there is no accounting for human nature or taste, and most people like to do just what they have done before, and the ancestors before them. I must stop one of the men are (sic) going to town, and will take this, he goes to buy ice and fresh meat and will be here by breakfast time tomorrow morning."

Hotel Cachipay at Bogota was a single story building surrounded by a barbed wire fence. American Heritage Center, University of Wyoming.

South American native helpers fill a
five-gallon petroleum tin from one of the
local seepages in the Magdalena Valley of
Colombia, South America, in October,
1916.   American Heritage Center,
University of Wyoming.

"An old well that the Wambersie Interests drilled at San Sebastian before the company commenced operations," wrote Kessack Duke White.   American Heritage Center, University of Wyoming.

Much of South America was still wild and
uncharted in 1916 when Kessack Duke
White investigated it.   American Heritage
Center, University of Wyoming.

**Kessack Duke White's expedition ready to
begin examination of seepages of the
Quebrada Verde on the Carare River
Expedition.   American Heritage Center,
University of Wyoming.**

On an expedition up the Opon River, the ox mired in the mud. K.D. White noted that it was a "very common experience." American Heritage Center, University of Wyoming.

"If I were a barrel of oil, comfortably located in a pool, hidden in a trap deep in the ground, the region that would be safest for me—where I might live out another 50 or 100 million years in peace and dignity—would be in some country where minerals and exploration are nationalized. The reason is that in countries such as these, there is but one hunter, and the chances of eluding him are far better than the chances of being discovered. The most dangerous place for me to live, as a barrel of oil, would be in the United States where there are thousands of hunters and each has a different weapon."

A.I. Levorsen

The next week, White was on the early-morning train bound for Colamar (Calamar). From there, he traveled via railroad, paddle boat and smaller boats up the Rio Magdalena to Honda at the first falls or rapids. "In Honda the owner of the Casa Inglesa and couple of haciendas was the nephew of Lloyd George. He had a prized stallion named Elegante. He was a beautifull horse, but a self opinionated one. If anything struck his legs he went 'berserk.'

"We were riding out to examine some outcrops. Hughes was not feeling too well and asked me if I cared to ride Elegante. I knew his reputation and had misgivings. I tried him out in the corral, put him through his paces every thing went fine so I rode him. It happened as we were climbing a rather steep slope and I was leaning far forward to take the weight off of the horse. I was using a McClellan saddle, they are slick as glass....A geological pick was in a saddle ring. The ring gave way, the pick fell striking Elegante on the legs. He bucked, I went over his head, landed on my face, breaking the small cheekbones below the eye. The Major Domo laughed, thought it a good joke." White was laid up the next day with a swollen face and black eye. "A month or so later I returned from the coast. I met this Major Domo Jose in the hotel, his arm was in a sling. I asked what happened, sadly he replied that he was riding Elegante, the lariat fouled and struck Elegante's legs and Jose landed in an orange tree." White broke out with laughter, but the major domo did not consider it funny.

By September, White had arrived in Bogota, "after having climed (sic) 8000 feet on mule back, as I took the trail to see the country instead of coming by train. It is very hard to believe with out having seen it that such a country exsists (sic). It is a plateau as level as the prairie of Illinois, with the mountain rising all around. Here only 250 miles north of the equator is a climate like autumn in the States where every one wears woolen cloths and an overcoat when you go out. It is a very delightfull country and in traveling thru it on the train from the fields of corn, wheat and potatoes, you would think that you were in the States. The city is not pretty. It has about 150,000 people, and reminds one more of parts of Naples than any thing else....As to the people, no where have I met them as delightfull. Every one has been either to Europe or the States and most of them to both, they are well educated and very cultured....They are in reality Spanish transplanted to America....I have met a number of people in different parts of the country who live here and they are making life very pleasant for me. I went last night to the club which is a very pretty place, excellently furnished with European furniture, and when you think of the great difficulty of transportation it must have run the cost up into the pictures. They have a Polo Club and I expect to go to the meet sunday. Of course this is all very well but it is a little hard on the work, so I only go out at night, though I am going out to Tea this afternoon. I must stop as it is morning, and in the mornings work must be done. I have just finished a six page letter

"The little girl which is the central figure... is a Cuban who came over on the boat with us from Colombia and is on her way to Bogota," wrote K.D. White. "She is a cute little trick. The young chap with the bag is a Mexican. The older fellow is the father of the girl." American Heritage Center, University of Wyoming.

For shelter at night, the crew built
"ranchos"--small houses or sleeping
havens. These were constructed in 1918,
at Contento on the trail to San Fernando,
on the Carare River.   American Heritage
Center, University of Wyoming.

to New York, and as the weekly mail leaves at noon today I am getting your letter off also.''

White's work was halted that fall when Standard Oil Company of New York abandoned its world-wide venture and recalled all parties from Colombia, Peru and China. The official reason was given as ''failure to discover reserves,'' but unsettled world conditions probably contributed to the decision. White did not let this stop him, however. He was back in Colombia the next year for the de Mares Concession. ''Leonard Hunley was the geologist for the de Mares people and we sailed for Cartegena on the same boat,'' White wrote. He mapped the Valanzuela concession, traveling up the Cachura (Cachira?) river in a dugout canoe.

''This type of boat, the most common in use, is constructed by hollowing out a tree trunk and shaping it in canoe form,'' he wrote. ''These boats generally are 15 to 20 feet long. They are paddled or poled. We were poling against the current along the bank. This river was really a channel in a swamp and had racing water as it was in flood. A drizzling rain was falling and everybody was miserable when a boatman, with his pole knocked a hornet's nest into the canoe. We were covered with hornets. I picked off a dozen and as a result of the stings had fever that night. While we were fighting hornets a boatman knocked a poisonous snake from the overhanging branches into the canoe. To add to our problem, the river had plenty of alligators.''

Another party in South America in 1915 was that of Joseph Theophilus Singewald, Jr., professor at Johns Hopkins University. Singewald explored the area that year in search of minerals for industrial wartime use. He and Professor B.L. Miller visited almost all the leading mines in South America. They paid their own way, but the trip was cut short by the war in Europe, which made conditions in South America unsafe for Americans. They returned to the states to work on a book on South American mineral deposits and dream of going back.

·EARTHQUAKES AND LAVA FLOWS·

While many of the geologists had been brought back closer to home to work during the world strife, Sidney Powers, a well educated young man, sailed for Hawaii early in 1915. The islands were still an exotic land then, and the natural geologic phenomena intrigued him as did the people.

''There is no telling how long I will remain here,'' Powers wrote his mother. ''I am going to try to get around this island in a machine, but it is a costly proposition. I have $155 left from my $500 and no return ticket to Frisco. Also my own money, which is enough to last me till May, if all goes well. The rate here is $90 a month, but I expect to stay here only a month. Tomorrow I shall talk to—about taking his machine to go among the island during his absence in Honolulu. Gasoline costs 40 cents a gallon at this hotel, 22 1/2 cents in Hilo....

"I am working on several problems, regarding ash and little cones. There are very peculiar pits here—holes 250' deep with a top diameter of 1/4-1/2 mile. One bottle-shaped hole is 35' in diameter at top, 250' deep, 250' diameter at bottom. Made by lava shattering off blocks; then lava runs off underground thru caves."

Two weeks later, still at the Volcano House, Kilalea, he wrote, "I am becoming anxious to see more of this vicinity than can be covered in a day's tramp, and expect to rent a machine by the week, or in some way, in order to get around the island....The lava lake remains at 470 ft. It should sink till March 24. It may disappear entirely at that time.

"Dr. Jaggar offered me a position here for 6 months, after June, as personal assistant; taking daily observations in Kilalea, doing the seismographic record-changing, taking either observations, tc; but the offer does not appeal to me, at about $90 a month with living expenses at $75, & doing merely clerical work which is worth only $1.50 a day in the states.

"I am favoring a trip to Japan, but that is still in the air. For 3 mo. (inc./mo. on water), the cost would be $600 acc. to Dr. Jaggar who went over last year. I get $300 from Harvard if I go, but the time proposition is the trouble."

Powers noted that water was a major problem on Kailua Island. "To run the mill, they have to pump up the water to shoot the cane down in sluice-ways. They would give anyone $50,000 or more for a permanent supply at the mills of 2 M gal. a day....If I had enough capital to set up a boreing machine, I think I could strike more or less water with little trouble. In the crater there are alternate beds of ash and porous lava. The water runs in the latter underground and comes out in large streams on the shore. The proposition is to locate an underground stream in the old crater....They employ all sorts of fakes to find water now and never employ a geologist."

·MEASURE BY THE FEET·

The Midwest was still drawing geologists in 1916. William G. Argabrite worked with the U.S.G.S. mapping in Kansas, Oklahoma and Texas. But conditions were not easy. There were still companies who were not convinced geology was the answer, but catered to the geologist because it was fashionable to do so. Sometimes a change of personnel could mean a complete turnaround overnight. Raymond F. Baker was working out of Tulsa for The Texas Company in 1916. "We were accustomed to return from the field every week end to get a decent meal and clean up," he explained. "On Sunday morning we would assemble in the office to discuss with Eckes our work, receive new assignments, supplies, etc. At that time a pompous ignoramus named Mike Connor had been recently appointed as the

Kessack Duke White poses with "Sr. Martinez" at Well 9, somewhere in South America. American Heritage Center, University of Wyoming.

White poses with others aboard a boat.
American Heritage Center, University of
Wyoming.

(Preceding page)
On the Opon River Expedition in 1918, the
entire group carried muzzle-loading
shotguns.   American Heritage Center,
University of Wyoming.

Remnants of an old shale still in Juab County, Utah in 1916. D.E. Winchester, U.S. Geological Survey.

Oklahoma division manager. One Sunday morning he paraded into the office where the geologists were assembled and declared himself somewhat as follows: "If I had the final say-so none of you fellows would be around here. You can best please me by getting the hell out of here into the brush and staying there."

William E. Wrather experienced the same negative attitude when he worked for Gulf. "Knowing I was a geologist, Gulf assigned me to the job in north Texas....I worked like the devil all over north Texas, and made various recommendations for drilling wells, but Gulf hadn't reached the point where they were willing to spend money on a geological recommendation. I went down to Beaumont and argued that I had a structure located up there in north Texas that justified a well, and I wanted authority to commit the company to leases and to drilling a well. They agreed to give me $20,000 for leasing only."

Wrather and a lease man lit out for north Texas and began leasing without keeping much track of the funds. A hurricane knocked out the phone lines; and by the time the lines were back up, the company was mad at Wrather for overextending himself. "They were not willing to spend money on drilling, following geological advice," he said. "I was all steamed up, wanted to find out whether my geology was any good."

Wallace Pratt encountered such an attitude with a division superintendent in Wichita Falls, Texas, a Mr. Little, "a dignified, quiet-spoken man, for whom I came soon to feel a warm affection as well as a high esteem," Pratt wrote. Little admitted him to his office when he arrived in Wichita Falls. "I stood in front of his desk. He did not rise. 'Mr. Pratt, you are our new geologist, Mr. Woodruff writes me. I wonder if you realize that we don't think much of geologists around here. Your office is at the end of the hall—the last door. When I want to see you, I will send for you.' "

Even when officials were favorable, conditions might be less so. In the spring of 1916, Allen I. Riley was instrument man for Alex McCoy. "In those days transportation was very difficult in parts of western Osage County (Oklahoma)," he noted, "and we had to depend on some of the scattered ranchers to provide us with a team and wagon for our moves from one township to another. Our living quarters were of the crudest; a tent pitched by a spring which provided drinking water, tables of packing boxes, cots for sleeping in the open, and our food was mostly canned."

By this time, every Osage Indian had become wealthy from his "headright." They bought fancy houses and thought nothing of driving their Cadillacs roughshod over pasture trails. Many a young geologist watched enviously as the Indians in their fancy cars passed by the footsore "poor white boys" who mapped the land and spotted the drilling sites that would bring the Indians even more riches.

According to Ed Owen, "The district chief had the only model-T Ford, so we walked out of town with our instruments on Monday mornings, stayed with the farmers for meals and lodging, and walked back to town on Saturdays. (Wilson C.) Giffin made regular rounds of the field parties to check our activities and give some needed instruction. Naturally our

work was of uneven quality, but we worked from sunup to sundown and learned rapidly; results on the whole were quite dependable."

Owen's original contract was with the subsidiary Continental Oil and Gas Company at a salary of $90 a month. "I was promoted after one month and received Robert B. Whitehead as my first instrument man. We enjoyed wonderful hospitality in the farming communities for Bob was an artist on any musical instrument at hand and was theatrically eloquent at the customary family prayers which preceded breakfast....

"Our eagerness compensated for lack of experience. The work was fun; the prospects were exciting; and nobody had any thought of 40-hour weeks, pensions or job security. Self-reliance was the way of life, and we admired even the stubbornness of Roy Butters, who thought that a straight line was the shortest distance between two points. Roy always drew a bead on his objective and went there directly, no matter what rivers, bluffs, or tanglewoods stood in the way. Our self-confidence undoubtedly made us difficult to manage. We were exploration geologists and were willing to explore anywhere for anybody. We were looking for a job when we found this one, and were always ready to look again."

By 1916, Empire was mapping geologic structures throughout Kansas, and the geological staff had grown to over 200 men. In June, the *Olathe Independent* noted, "Olathe has a life-size mystery on its hands and so far, not even the Olathe senate has made any progress toward a solution.

"A band of men dressed in Khaki and equipped with surveying outfits have been operating in and around Olathe for two weeks or more.

"Eight or nine of them have been making headquarters at the National Hotel and this week seven more came in.

"They are skilled at not giving out information, and all questioners get a cool reception. Here are the theories which have been advanced:

"1. They are railroad surveyors laying out a new railroad with Olathe as an important junction point and speculators are busy picking up options on all valuable tracts of land.

"2. They are German spies laying out military roads so they can most easily reach Kansas City, the heart of the continent, from either coast or the Gulf of Mexico.

"3. They are emmisaries (sic) of the Japs looking for an aerial base for safe landing when they jump into the Mexican conflict against the United States.

"4. They are employes of a big oil company making a geological survey of the county to locate the immense pools of oil which underlay Johnson county.

"5. They are government surveyors laying out locations for great munitions plants which will make Olathe the great military manufacturing center of the United States.

"6. They are a band of engineering students out for field practice.

"There are several other theories afloat but these are the ones that have been booked. Every citizen is entitled to a guess of his own, however, or if he wants to try his skill at questioning, the way is open."

The technical staff of the Division of Mines, Bureau of Science, Government of the Philippine Islands, in June, 1911. Back row, left to right: F.A. Dalburg, graduate of the University of Western Pennsylvania; Frank Eddingfield, mining engineer; Warren DuPre Smith, chief of the division of mines; Wallace Pratt; unidentified Filipino mining engineer; Andrew Rowley, petrographer trained at University of Chicago. Front: three Filipino staff members. In the center was Gregorio, laboratory handyman, and to his right, "our lithographer, a talented artist." Not shown in the photo were Paul Fanning, metallurgist who was trained at MIT; George I. Adams, senior geologist; and Henry G. Ferguson, hard-rock geologist who returned to the states and was employed by the U.S.G.S. exploring for gold and silver in Nevada for the balance of his professional career. Wallace Pratt/AAPG.

Amerada had Sidney Powers traipsing around Ardmore, Duncan, Healdton, Hewitt and Loco, Oklahoma, searching for promising structures. By this time it was a lucrative area, for the Healdton field was producing 90,000 barrels daily in January 1916. Powers's letters to his mother revealed that geologists really did think about things other than geology. "The enclosed circular will be of great interest to the bloated bondholders at the CRW who are making a fortune in oil," he wrote from Ardmore, October 27, 1916. "Was the last appointee to the Supreme Court, the man who kept Wilson & his wife from separating? What is the scandle in Wilson's life?...

"Sorry you wrote E. Louise about girls in Shreveport. One girl is engaged; the other never answered my letter. I have no time for society...They have 6 mo. work all laid out ahead in the field.

"EG Woodruff is coming tomorrow to see if we are able to work out geology. He either thinks that we are the kindergarten class or else that there is really oil here. There is no structure in our area. S. Oklahoma is a...high rolling, open country with occasional small patches of low oak trees separated by vast expanses of cultivated land. (hay, cotton) Houses scattered, small & one story. Always a dug out hurricane house nearby so that the family can all crawl in when the wind blows. The houses differ from E Texas in having windows of a more solid construction but they are separated from the ground by 1'-2' so that they are easily removed by the wind.

"Ardmore is a rapidly growing town of 12,000 + with one or two trolley cars, several movies, etc. No drinks are sold in old Indian Territory under heavy fine, & even liquor in suitcases is confiscated on account of the Indians who are scattered around this country...

"At Healdton, Loco, and Fox they have oil & gas fields (15 m + W of here). Hundreds of autos of all kinds, but mostly new, travel incessantly over the road thither & everyone in this country knows us as we spent 2 days along the road with a team of small horses and a spare tack....

"Leases in Oklahoma cost $5-50 an acre; in Texas 10-25 cents. Everyone has the oil craze. They are leasing land in Nebraska now—Kansas largely leased."

On November 12 Powers wrote, "Dear Mere, You need not count on my being home Xmas....I had the Johnson pants repaired & wore them today. At the first fence the strain was too great & they ripped. As soon as my other pair is repaired, I shall send these back to Johnson....Roads are fair in this country, all on section lines, but when one gets into the southwestn part..the roads are better suited to horseback or on foot than anything else—sandy & full of stumps. We got stuck for awhile yesterday when the ruts became too deep.

Powers found himself in the middle of a Midwestern storm one afternoon. "The cyclone came up in a hurry, with dense black clouds, and just as we got the Ford behind a barn a terrific sandstorm arrived. It blew many sheet-iron slabs off the roof of barn & butted the

Wallace Pratt poses in front of a 16-foot coal bed near Liguan, Cebu, Philippines, circa 1912. When Pratt identified the photograph, he noted--like any good geologist--the formation behind rather than the people in the foreground. "Shattered outcrop of a sub-bituminous coal bed 16 feet thick tilted to an angle of 45 degrees at the barrio Liguan on the axis of the Island of Cebu," he wrote. His "Filipine capitas," better known as "Ting," exposed the film. AAPG.

farm wagon into barn. The inhabitants of the house disappeared in the cyclone house, but reappeared 2 minutes later when the cyclone passed & a 20 (degrees) cooler N wind started. It was the most sudden drop in temperature I ever experienced."

The team put up in the Healdton area, in a two-room house with bath for $2.50 a day. "E G Woodruff will undoubtedly write a notegram on the subject," Powers wrote. "He wants us to sleep in a 25-cent house. He wrote Milyko in July congratulating him for living in the YMCA at 75 cents a night & living so cheaply (meals averaging 35-40 cents), and on the same day wrote (Lewines? Lemines?), an American, criticizing him because he tipped the Pullman Porter 50 cents a day."

Powers complained also about the prices of canned goods on which they had to live. "Van Camps beans 2 for 25 to 20 cents straight; canned herring 35-50 cents. Herring from the N Sea should taste familiar to Americans this year as they have eaten so many of our countrymen. Canned peaches 30-35 cents. We ate beans, sardines & apricots (canned) on Thus & and all three fought in both our stomachs all afternoon. We still survive."

Sidney Powers and his party crossed parts of Europe and Asia in December 1914. American Heritage Center, University of Wyoming.

Chester Longwell on Mount Guyout, in the winter of 1915-16. E.W. Owen/Mirva Owen.

·PIE FOR BREAKFAST·

In 1916, Sidney Powers was in Texas looking for potential drilling and lease sites. Powers was intrigued by southern habits. In April, he wrote his mother from near Shelbyville, "It is difficult to enumerate the features of this country which differ from any other place I have been in.

"The houses are usually built in 2 pieces so that wind blows thru.

"The women milk the cows & do most of the work. The men are infinately (sic) lazy.

"Ploughing with 1 horse & a tiny plow is carried on everywhere for corn & cotton. They plow in such a manner as to leave the remains of last year's crop not turned over. Occasionally they run a tooth harrow over the strips 2'-3' wide between the plowshares....

"Pigs are everywhere, in the woods, the roads, etc. Some are the razor-backed variety & never have had a real meal. Very few of them are ever fed....

"No bread is baked; only baking-powder biscuits. Fried Eggs are on tap. Occasionally a chicken or a goat, tho the latter are rare. Corn & corn bread are always available.

"Breakf. 6.15-6.30; Dinner 6.30-7.15. Lunch of biscuits, fried eggs, a strange kind of apple sauce in the form of pie, & cake comprises the noonday repast. The water all comes from wells & is usually slightly clayey. $1.00 a day. Room & Board."

On September 9, he wrote, "I have been intending to send you a can of snuf & a snuf stick for some time past so that your gums could enjoy the pleasing sensation that captivates East Tex. women. Yesterday I saw a girl of 10 with a snuff stick in her mouth."

Some companies—and their geologists—were more successful than others in the Texas country. Producers Oil Company kept trying to locate oil, but one disgusted geologist turned his files over to young Wallace Pratt and remarked, "Why do you waste your time in Texas? Why don't you come up to Oklahoma where the oil is?"

·ANYONE FOR SCRAMBLED EGGS?·

In 1917, Ben C. Belt moved to Ranger, Texas, working for Gulf Oil. Ranger was just getting started then: the first well had just come in. The first day on the job, Belt went out to look at a well and when he returned to town, some enterprising soul had set up a cook shack. Belt was fairly hungry at the time, but he could think of nothing that sounded particularly good. He finally settled on two poached eggs. He waited—and waited.

Finally, the cook stuck his head through the window where he passed out the food and hollered at his partner, "I told you those eggs wouldn't stand poaching!" Needless to say, Belt passed on the eggs.

(Opposite)
Men and women view geological strata in a flooded underground mine. George Hansen Collection, Brigham Young University.

(Right)
Oil Shale Mining Company's first retort on Dry Fork west of De Beque, Garfield County, Colorado, in 1917. D.E. Winchester, U.S. Geological Survey.

*"Prospecting for oil is a dynamic art—
a series of individual techniques—sometimes
overlapping, sometimes separated by a time gap—but
techniques which lose their usefulness and which
leave us without guide to our prospecting until we
devise some new method. The greatest single element
in all prospecting, past, present, and future—is the
man willing to take a chance."*

Everett DeGolyer
quoted in Mr. De

In the winter of 1916, Horace Griley was recruited for work with the Rocky Mountain Division of Empire Gas and Fuel. "Glen Ruby obtained my signature on an employment contract," Griley wrote Ed Owen. Ruby was then chief for the Rocky Mountain area, with offices at Billings, Montana. "I agreed to work for one hundred dollars per month plus field expenses. During the summer of 1917 in Wyoming I learned that some men with no more than four years' university training were getting $125 and $150 per month. I was too hesitant to ask for more than the one hundred payment. I went to National Refining Company (Frank Carney, Chief) for $125. "(Ruby) had a field party in southern Montana...one or two parties in Colorado (and) about 15 parties in Wyoming. For the July Fourth holiday, 1917, all men working under Ruby were brought to Thermopolis....We filled the frame hotel. The number was in the forties, perhaps 42....It was probably true that the total number of men in Empire's field organization, summer of 1917, was about 250. They were in New York state (Jimmy Thompson), Kentucky, Texas, Ohio, and just about everywhere, if you count prospective oil territory of that day and time. Some of the men who were with me in Wyoming had been in Cuba the previous winter."

Camp life in the field could be strenuous. "We lived in tents, set at the common corner of four townships. After walking all day we sometimes quit about 4:30 p.m., approximately six miles from the two-man camp. The tents were designed for simplicity so that hauling (one) in a Model T Ford after folding was not cumbersome. There were two short tent poles at the front corners, a long sloping back, hanging sides, and a large front 'fly.' Until about July 10th, when night temperatures moderated we slept on bed rolls placed on the ground. Later in the summer we used army cots. The space in the Ford car between front an(d) back seats held a wooden grocery box. The car was driven by a high school boy who came to the camp sites of a field district about twice a month. The men at a tent location might be taken to town for a day or two, principally to buy groceries. An area chief would come out when necessary to supervise a move to four more townships. Cooking was over a fire of dead sage brush roots, inside a triangle of rocks. The exercise in the mountain air produced good appetites." Since lunch had to be confined to what fit in a man's pockets, it was usually light. But the men made up for any deficiency at supper.

"Practice on extremely light to extremely large meals each 24 hours enabled me to put on a stunt in the Burlington dining car when I was moved from Cody to Lander. First I consumed a full noon dinner. Then one after another I ordered and ate all meat and fish courses listed on the menu. For an hour after the car was closed the waiters stood around watching me consume food..on a company expense account."

Empire's geological mapping in Wyoming evidently did not pay out compared to the cost. On September 1, Empire cut out more than 50% of its field geologists. Only a few were transferred from the Rocky Mountain division to the Mid-continent. Griley remained with Empire but moved to Oklahoma.

"Only a Flivver" on its way to Ponca City--and like most cars of its day, it broke down before the trip was over. Courtesy, Robert Dott, Sr.

In addition to mudholes, there were always flat tires to plague those who preferred the new-fangled transportation. Courtesy, Robert Dott, Sr.

Cars replaced many of the horse-drawn
vehicles, but the horse team was not
totally obsolete. When the auto ran into
trouble, it was the reliable horse that got it
out.   Courtesy, Robert Dott, Sr.

In 1910, C.J. Hares was working under Carol Wegemann around Salt Creek, Wyoming, for the U.S.G.S. He was sent in 1911 to map lignite in North Dakota (shown "near Marmarth") and returned in 1913 to map structures in central Wyoming. His assistants were Max Ball, Stuart St. Clair, J.B. Reeside, K.C. Heald and A.C. Collins. Most of their work was on foot or horseback.   Courtesy, E.K. Erickson.

·DRAINING THE AUTO·

Just after the Ranger, Texas, boom in 1917, Ben C. Belt moved to Eastland where they had "the best hotel in the county" and a firm sheriff that kept the town honest. "No one worried about petty thievery, even in the boom times," he said. "You could leave your keys in your car, even your suitcase."

Water was another matter. There was a drouth that year, and "any man who didn't carry around a big bucket so he could drain his radiator every night was a real greenhorn. It was standard procedure every time you parked your car to drain it and carry the water up to your hotel room."

·DANCING TO THE TUNE OF OIL·

Another veteran of the Wyoming oil patch was Charles Hares. "In 1917, Charlie Hares was out in the field and came upon an oil camp whose crew chief invited him for a cup of coffee at the camp fire and told him they had drilled two dry holes and were starting a third test," Eric Ericson wrote. "Charlie knew the country around there pretty good and told the chief the third test would be dry too. A few months later he was proved right and the chief told his Ohio Oil Company boss John McFadyen about Charlie Hares prediction." McFadyen supposedly blurted out, "Who the devil is Charlie Hares?" "Ohio didn't have any geologists," Ericson wrote, "but McFadyen figured it was time and convinced the president, J.C. Donnell, (who had previously stated he'd never hire a geologist) to give Charlie a chance." Ohio made Hares an offer. "After some trading, a meeting of minds was arrived at. In two weeks, he left the Survey...He sensed that if he did not accept quickly that company would hire some other geologist. Anyway, his salary of $2,000 barely paid the rent and grocery bills in Washington...He perceived a real opportunity to apply practical geology in the search for oil and gas. Morever, if he could not make a success of it, he just might as well go back to hoeing corn with which he was familiar, on the glacial drift of New York."

Although Ohio had occasionally hired a consulting geologist, they had never made a discovery on one's recommendation so Hares regarded his employment as a challenge to prove the value of the oil geologist. In those days, it was a common saying in the oil fields that if you wanted to be sure of drilling a dry hole, you only had to hire yourself a geologist. Hares was out to prove the saying wrong.

By this time, Casper was dancing to the tune of oil. "The oil boom in 1917 was on in exciting fashion," according to Hares. The town overflowed with the usual boomtown businesses. It was said that there were not enough sandbars in the North Platte to hold the joints opened up to separate the ever hopeful oil men from their money. Hotel lobbies and streets "were full of milling women and men, while the beds had a good supply of bed

bugs,'' wrote Hares. "Hotel rooms were at a premium, if a room had a double bed two persons were assigned to the room even if they were utter strangers. A smart chap could register in a dummy then pay double.

"New oil stocks were traded freely in the lobby of the Midwest Hotel (later the Henning). A chap by the name of Blaz remarked that a stock he bought on one side of the lobby was down several points before he could walk to the other side and sell it. Blue chip stocks of the N.Y. Stock Exchange could be dealt in in Casper as well as the 'cats and dogs' in the bucket shops.

"In 1917, the Ohio Oil Company had only two rooms for an office over Campbell and Johnson Clothing store in Casper, and 4 persons occupied them, McFadyen, Holland, John Marvin and myself."

Hares's report, USGS Bulletin 641, covered the anticlines of central Wyoming. It was every oil man's bible—laid by the bed, carried in his briefcase or stuffed in a pocket. "Charlie's bulletin...was a best seller around Casper in those days," wrote Eric K. Ericson, "and the new discoveries didn't do Charlie's reputation or confidence any harm."

Rotary rigs were growing more popular in 1918. Esperson's rotary rig was located at Wynnewood, Oklahoma in June 1918. Driller was John Wade.   Courtesy, Mirva C. Owen.

C. J. Hares gets ready for a day's work in the field. Note the stovepipe in the tent. Courtesy, E.K. Erickson.

(Above, left)
Ed Owen at the wheel. Courtesy, Mirva C. Owen.

(Below, left)
Frank Gouin carried his Empire 15-foot rod on the side of his Model T. It was a tremendous improvement after the horse-and-buggy days when the surface party had to leg it almost every foot of the way from shot to shot.

(Opposite)
Harry F. Sinclair and John Overfield in 1907, in an early electric car. Gasoline-powered vehicles would take over the field within the next few years, as Ford released its Model T. Atlantic Richfield Archives.

At that time, the Midwest Oil Company, the Midwest Refining Company and The Ohio Oil Company were the main oil companies doing most of the drilling and the acquisition of properties. "Henry M. Backmer told me that the reason he organized the Midwest Refining Company was because the OPC titles in Salt Creek were in jeopardy and as long as he could buy the oil for the refinery he cared not who had the oil for sale," Hares said. The Ohio Oil Company would not gamble on buying the risky titles of the oil claims in Salt Creek. But there were plenty of less active oil companies—Texas, Gypsie, City Service, G.P., Union. New oil companies were formed almost daily such as the Utah Oil and Refining, N.Y. Oil, Kinney Coastal, McKinney, P and R, Bishop, Bair Oil, Geoughan and Hurst, Conroy and many others, some of which amounted to very little.

"The Stock boys were omnipresent," Hares wrote. "Practical oil men in those days had the habit of poking fun at geologists, called them ridge runners with no business sense, also noted they dressed queerly in khaki and puttees, were easily spotted in hotel lobbies and on the streets." Hares wanted more anonymity to do his field work, so he adopted a business suit and "dancing slippers as some of the boys called my soft kangaroo leather shoes which were (are still) non-peeling even among rocks. They did

Another breakdown. Sometimes geologists yearned for horses, who were at least no more fickle than the flivver. Courtesy, Robert Dott, Sr.

not tire me and what was more important they left the hob nailed boots of other geologists behind on the ridges and in the sage brush.''

Ericson noted, "Scouts would follow Charlie around and he used to disguise himself with office clothes just to get away to the field. Parker Trask of Shell Oil once claimed that his job "was to watch Charlie Hares and follow where he was working."

During Hares's first year with Ohio, he made three important discoveries—Lance Creek (giant), Rock River and Hidden Dome. The Ohio was convinced that geology had its merits. "The practical oil men even if they razzed geologists did have a lot of respect and use for them," Hares wrote. But he did not add, as others did, it was because of Hares's own success. "Uncle Jack as we called Jno McFadyen then manager for The Ohio and an

Empire Gas & Fuel Co. had geologists running all over the Kansas area doing detail mapping in 1916. Most of them were afoot, however, as there was only one auto for the party of 30 geologists. Left to right: Calvin Moore, Wilson Giffin, Mason Read and Donald Barton. E.W. Owen/AAPG.

indefatigable worker, wanted me to teach all the hands geology," Hares said. "When I asked when, he replied 'Sundays and holidays'. Being too busy for such a silly job no class was formed. Besides it seemed a mighty poor idea to teach my stock-in-trade to anyone. Let them dress tools, mess around in the laws and drive trucks....

"Frequently, upon coming to the Casper office McFadyen would ask me, 'How soon can you look over this acreage that Pat Sullivan, Gene McCarthy (or some banker) has put up to us to drill?' The only satisfactory answer was as soon as the car is greased or the Burlington train leaves."

Between 1917 and 1921, Hares claimed he never made a contour map nor a geologic map of any prospect because there was no time. "Virtually all of my turn downs and favorable recommendations were made by long distance telephone or verbally in the Casper office. Bankers, ranchers, promoters and who(ever) were constantly putting up stuff to the management to drill which out of good public relations had to be looked over generally in the field. This system certainly gave freedom of action and got results fast."

Remnants of an old shale still in Juab County, Utah, in 1916. D.E. Winchester, U.S. Geological Survey.

·STAGE COACH TO CANANEA·

When young graduate Lewis Weeks was invited to work for Greene Cananea Copper Mining Company in Sonora, Mexico, in 1917, the work sounded good to him. "I had never been farther from home than Chicago or Minneapolis but I was thrilled at the prospect of traveling thousands of miles to a new country. Besides, the pay, $90 a month, was good for a beginner....A battered steamer trunk held my worldly possessions, with room to spare, and the Chicago, Rock Island and Pacific Railroad carried it and me in three days of solid traveling to the small border town of Naco, Arizona. In those days, the trains, of course, lacked air conditioning. In warm weather, you just opened the windows of the railroad coach and the smoke and soot came in with the fresh air.

"I arrived in Naco around 11 at night, dirty and hungry, and the train pulled away leaving me utterly alone on the platform. A few minutes later, a messenger appeared with word that the town was a long hike away and that it had been attacked by bandits from across the Mexican border only a few days ago. The messenger offered to take one end of my trunk, while I carried the other, and as we walked to Naco, he advised me that I should put up in one lodging house that was made of adobe and turned a blank wall to the border. Before I checked in, I looked at the Mexican side of the building. Sure enough, it was pockmarked with bullets. But none, apparently, had come through."

The next morning, Weeks hunted up the border agent for bound for Cananea sixty miles to the southwest. The narrow gauge train was not running. Traveling by stagecoach over a dusty unpaved desert road was no hardship for a farm boy like me and the journey was quite like the stagecoach scenes shown on western movies. Repeatedly during the day we were halted by the mounted police called "Rurales," who scoured the countryside under the direction of the military government. Fearful of new insurrections, the military had issued orders to challenge everything that moved. Each time we stopped, we had to pile our baggage on the ground, open it and let the police examine it to be sure we were not carrying weapons or other contraband.

"We finally reached the company's headquarters camp late in the day and I was assigned to a bunkhouse and then invited to sit down in the company dining room to an evening meal prepared by the Chinese cook. In those days, the Chinese were a fixture of mining camps. In nearby towns they were the principal merchants; in the camps themselves they filled the need for servants. Over a hearty dinner, I learned the reason I had been summoned in such haste. My boss, the company's chief geologist, was getting ready to leave to return to the States and I was expected to step immediately into his shoes. My duties would consist of surveying, sampling ores and recommending where to sink new shafts or laterals, plus maintaining up-to-date maps of all the mine workings." But as luck would have it, Weeks had arrived in Mexico just as Pancho Villa's guerrillas were battling their way through the countryside. "The hostility of the troublemakers to gringos and to mining interests owned by North Americans was no secret," Weeks wrote. "Within two weeks of my arrival, a message came to our camp that all gringos should clear out before nightfall. There happened to be a train running that day to the border and when it pulled out, I was aboard. I later heard that the guerrillas arrived at dusk, thoroughly looted the mine office and forced the closing of the enterprise."

The Cimarron River ran through the middle of the prolific Cushing Field, in Oklahoma, and the best way across was by ferry. The well at the extreme right was the first one brought in on the banks of the Cimarron. Magnolia Oil News.

''I regard geology as a cultural subject, as much so as art or literature. In the dictionary sense, it contributes to the 'training and refining of the mind.' It does so, in particular, by making us aware of the immense sweep of geologic time and of the vast forces operating in the earth as we see it today.''

Lewis Weeks

The U.S. Geological Survey camped near the head of Battlement Creek, 10 miles south of Grand Valley in Garfield County, Colorado, in September 1918. D.E. Winchester, U.S. Geological Survey.

The search for oil grew more heated along with the political problems. On May 4, 1917, Sidney Powers sent a succinct mailgram to his family: "Sail New Orleans Seventeenth Tela Honduras with assistant two scouts and secretary company going sooner examine north-eastern portion country only three months then one month vacation all with three times present salary send things shreveport requested yesterday telegram. Sidney."

He sailed on a United Fruit Company steamship, arriving in Belize May 21, 1917. "The industry of the colony is mahogany shipped from the Belize River which flows thru town," he wrote his mother. "The river is led thru the town in small canals which furnish a convenient site for outhouses. Where there are no canals, open smelly sewers retain all the contents which mold in the heat. Back of town are innumerable puddles of all sizes, supporting a freshwater life of mosquitoes, etc. Malaria is extremely prevalent and mosquitoes & sand flies bother the natives. Water is promised by the Lord only, but the supply is adequate...Dress is European. Shoes usually superfluous. Liquor dispensed in quantities except on Sunday. No work on Sunday, as is typical English colony. They take mail & passengers only off steamer....Business conducted by Englishmen who live over their shop until they acquire sufficient money to go home. All employees are of a yellow (jaundiced) hue. Hours: black tea 6 am; business till 9 am; breakfast 9-10 (all shops closed); business 10-4; light lunch noon; supper 7 pm. Polo, cricket, etc. ad interim. Bathing at the keys in protected places only on account of sharks."

From Belize, he set sail to (Puerto) Barrios. "No railroads in B. or Spanish Honduras of any moment—none in the former except possibly at the forests," he wrote. "Tegucigalpa the capital of Spanish Honduras is reached by hard horseback trip over the mts in 7 days, eat up tortillas (roasted cracked corn cakes, like pancakes, which form the solitary & national food)...."

"Panzos is nearly a mile from the place where the boat ties up, at which point there is a Co. building, shed for 2 RR cars, house for German bookkeeper. Walking up the track one paddles thru a swampy walk to a scattering of white houses. One of these houses is reserved for travellers. No food is available there & one must go around town & see if any woman will be kind enough to give one tortillas, frijoles & fried eggs. The sleeping house is not screened—except the end inhabited by the mulatto US Engineer and his negro wife. Millions of mosquitoes inhabit the house and furnish a serenade at all hours of the day and night. The Anopheles, of which there are many, do not sing. Upon crawling on the cot beneath ancient mosquito bar, dozens of the delightful pests at once roost on the bar and nibble holes thru it. Holes already there furnish access for those familiar with the travel routes, and soon one feels a bite & finds 6 bugs already inside. After half an hour's perspiring another bite and a large flock is found."

In Barrios, Powers began having personnel problems. "E—has had what he calls a good time the last 2 nights & has spent all his money. The result is that he had not enough money to pay his hotel bill or to buy a railroad ticket last night & I did not have enough extra money to fix him up—I lacked $5 so he is spending today in (Guatemala) City cashing a draft to get enough money to come to P. Barrios. His usefullness on this trip is great only unto himself. He greatly enjoys talking by the hour with anyone who has the power of speech & drinking beer with any old bum. Aside from these occupations, sleeping comes in for a large amount of time. Geological information microscopic. Far better suited to coal miner or sewer digger than anything else."

By June 25, the country was at war with Germany. "All bridges are guarded by uniformless barefooted soldiers armed with single shooters," he wrote. "One cannot walk the RR track...Salvador and Mexico harbor all the Germans who are aggressive. Most of the coffee plantations of Guatemala are owned by Germans. The crop was formerly shipped to England and Germany but England refuses to take German coffee so the price has dropped and the country is stagnating."

Charles G. Carlson (right), and his assistant, Walter Hayman, encountered many unusual natives--including a boa constrictor--during their exploration of the Orinoco River in eastern Venezuela in 1917-18. It was Carlson's first job, working for Aluminum Co. of America. AAPG.

"Guatemala is ruled by Cabrerra (who) lives in seclusion in G. City and directs his ministers what they shall do, and directs his secret agents who they shall kill...Cabrerra has been bombed, dynamited, shot at, almost poisoned, stabbed, etc., many times. Once the horses and coachmen were killed by a bomb from the top of a building; again when he stepped into a carriage it blew up...The Barrios Commandante killed the British Consul a few years ago...At night policemen are afraid, so their place is taken by soldiers with gun & bayonet who crouch in groups of 5 or 6 with a commander, on occasional street corners."

Powers did not tarry with his assignment. On July 4, he wrote from Livingston, "I will be en route home before you receive this. We are all thru but the shouting & cannot celebrate today as our guns are in the custom house."

Kessack Duke White poses for the camera in Angola, in full regalia. American Heritage Center, University of Wyoming.

·BEYOND SPANISH TRAILS·

In early autumn of 1917, Kessack Duke White was hired by one of the companies of the Brady interests to explore and evaluate petroleum prospects of the Upper Magdalena River Valley of Colombia, South America. War had been declared, however, and White, who was subject to the draft, had difficulty getting out of the United States. Finally, friends in high places helped him get clearance to South America. White headed out on an expedition into the headwaters of the Carare and Opon rivers. This area was reportedly dangerous Indian country.

After a month waiting for the rains to stop and the trails to dry, White pulled out of Bogota, December 27, on a narrow-gauge railroad for Nemocon, northernmost town of the Sabana de Bogota. It was a frustrating trip. "On the way the train crew mixed up my baggage and threw my saddle off at a wayside station," he wrote B.W. Dudley in New York. "That evening after some difficulty I secured three pack animals and one saddle animal to Chiquinquira, by paying practically double the regular fare. This occurred notwithstanding that an agent was supposed to have made all arrangements for the animals."

On December 28, he started the cook with the cargo for Chiquinquira. "The animals were supposed to be loaded by seven," he wrote. "They really left town at nine thirty. I was to follow when my saddle arrived, which had been promised without fail to come on the ten thirty train. The train arrived without the saddle but by much telephoning and telegraphing it was finally located and sent on a hand-car from a station about ten miles down the line to Nemocon."

White reached Chiquinquira, a city of about 15,000 inhabitants, December 29. "It is a very old town built after the fashion of the better towns of the cold country. The hotels, however, are very poor and very dirty, and inhabited by that ever-present insect the flea, which is so common in the houses of the colder climate."

They hunted all afternoon for animals to go on to Valez but could secure none before

Monday morning. They did not get away until mid-morning, hampered by a muledriver who had drunk too much the night before.

Valez was the oldest settlement in the interior of Colombia, built as a collecting point for European imports to the Sabana in the days of Spanish domination. White's entourage arrived about noon of the New Year during the local celebration. It took twelve days to secure guides and animals for the trip into the headwaters of the Carare River. It was not, however, an unpleasant wait. "I had letters of introduction to a number of people in the town," he wrote Dudley. "Tuesday afternoon was spent in meeting the people....It soon became apparent that the difficulty was going to be to get rid of the number of kind offers, rather than to secure assistance in obtaining a guide, men, animals etc."

On January 2, White sent a man for a well-known and experienced guide, Pacho Herrara, with orders to bring animals for the trip to the Carare. "It appears that the trail from Valez to the Carare is unusually bad, the mud is deep and frequently shelving rocks that slope off to the canyon below have to be crossed," he wrote. "Unless the animal is very accustomed to the route he will not only break his own neck but the rider's as well. Also there are bad fords in the large stream that has to be crossed several times. During the wet season this stream is frequently impassable for days at a time." White waited. Herrara did not show for several days. The rains, however, continued and conditions worsened. White had little choice but to head out for Valez—it was the only way to enter the region.

White detailed all his difficulties for good reason. "I have written you at length of the trip...so that you may know of the delays that occur even in the most populous and accessible part of Colombia, and will more readily understand how it is that work that should apparently be completed in one week may drag itself into two or even three weeks. Also, how it happens that even with the rush to finish a very hurried reconnoissance (sic) of this region it will probably require until the end of February, and time may string itself out until the first weeks of March. Nothing can be told definitely, the rise of a stream over night may prevent passing for two or even four days. The Rio Horta is such a stream, and it must be crossed to reach the Carare region."

On January 12, accompanied by Herrara, his cook, mule driver, cargo and saddle animals, White left Valez for the Carare River. The trail from Valez climbed the watershed between the Suarez and Carare rivers and crossed the range named Pena de Valez at a place called Guachanaqui. From there, they commenced their descent into the Carare River drainage basin. They had little trouble on the steep trail until just before reaching Racine. There they encountered the paved part of the old Spanish trail. From that point, progress was slow. "The reports of the trail, of which I wrote you previously, hardly did it justice," he wrote B.W. Dudley. "It is the worst and most dangerous trail I have ever seen. The trail was built in early colonial times by the Spanish, and since they abandoned the use of it no repairs have been made. It was originally paved with limestone slabs, which with the wear of time have become as slick as glass. It is impossible for an animal not accustomed to the

> *"We usually find oil in new places with old ideas. Sometimes, also, we find oil in an old place with a new idea, but we seldom find much oil in an old place with an old idea. Several times in the past we have thought we were running out of oil whereas actually we were only running out of ideas."*
>
> *Parke A. Dickey*

stones to stand. The trail also has many very steep assents (sic) and descents. My mule fell twice, once on top of me. The first time I was riding an excellent little animal, very sure footed, she put her hoof in a crevice between two stones and it caught, I attempted to dismount but my spur became entangled in the stirrup so that, when the mule fell it draged (sic) me under her, fortunately the ground was soft so I suffered no bad effects. The mule however, sprained its leg so that I had to change animals. When the second animal fell I was able to dismount before it went down."

One animal was left at a house by the road side. "The trail on the south side of the Rio Horta before reaching Landasuri was exceedingly hard going...When the mules, which are very intelligent animals and knew the trail, reached the ford of the Rio Horta they balked and it was with great difficulty that we could get them over. After making the ascent it was easy to realize how fully the fear of the animals was justified in attempting that muddy, dangerous piece of trail."

They arrived in Marcella January 15. From there they moved into the watershed between the Carare and Opon rivers and on to the Carare River. "On January 17th accompanied by four pack men, a guide and a cook I left for the trip to the Quebrada Verde...The Carare River region is a section little known, and for that reason and the additional fact that the Spaniards used it as a highway it is the source of many tales of wealth and richness. Endeavors have never been lacking to reopen it as a highway, and many miles of graded road now overgrown and abandoned have been constructed. A telephone line was once started to the Magdalena River, and an American company built a settlement to collect ivory nuts but all has been abandoned and the region is now practically unpopulated, and can only be entered by a trail that has almost grown shut."

White was back in Valez again on the 30th but set out with three cargo animals, an *arriero* (mule driver), a cook, and one of the representatives of the property owners. "We reached LaPaz late that night in a thunder storm," White wrote. The following day, the cargo was divided between two oxen, a mule and four *muleteros* (pack men) and they set off on the trail across the watershed down into the drainage of the Opon River. "The trail, though it hardly deserved that name for the first day, was difficult travelling. I had a saddle mule which I was supposed to be able to ride, but which really only carried the saddle except for about one hour during the day."

On the afternoon of February 5, the group arrived at the "little collection of clearings called Opon." From there they traveled to Sardines, the last habitation on the headwaters of the Opon River. "No people except wild Indians live in the region below for the entire distance down the valley to the mouth at the Magdalena River," he wrote. "The head man is the guide who picks out the trail doing very little cutting. The second man is a *machetero* who is supposed to clear the trail sufficient to allow the cargo men to pass without difficulty." All the men were armed. "The region we were entering was dangerous and it was necessary to go prepared to protect ourselves in case we were attacked.

"Travel in the jungle generally end(s) about 4 p.m.," White explained. "That is the time, according to local belief, legend etc. that poisonous snakes begin to move and travel....

"Making camp, in uncharted jungles, especially where it rains a bit every night, consists of building a low roof rancho of poles and broad leaf fonds (sic) and covering the ground also with fonds on which to sleep. For me they built a shelf above the ground on which to lie. Your bed roll was a blanket. It is different when mounted and there are no nightly rains. Then one clears a place around the camp fire, spreads his saddle blankets and uses his poncho for a cover.

"We packed in by mule and oxen to the last settlement where it was reported the Indians, a few years earlier, had killed a rancher. We had a dozen or so men who carried a pack and a muzzle loading gun. There were three guides who scouted for the path and cleared it with machetes. My control was by pacing and compass. Even though crude this is a very good and quite accurate method of mapping. The evidence of the trail was very poor, a rut or a scarred bank but even with the twists and turns it followed a general bearing. Then we began to drift off that general trend. I called the guides' attention, but they maintained we were on the trail and they knew where we were.

"I platted up my bearings and paces so had a fair idea of our general position. We continued cutting trail and traveling the next day and by evening reached a large stream or river, apparently the Rio Carare. We had completely bypassed and missed the Rio Opon drainage. We wandered around for another day. At camp time the guides admitted they were completely lost and a little frightened. We were man packing, and had too few supplies to risk trying to cut trail for the Rio Magdalena."

After three days aimlessly wandering about the jungle, White took matters into his own hands. "I had my pace traverse and compass map," White explained in a letter. "Our last identifiable point was a small pond on top of a hill. We had about a quarter of a mile leaway (sic) to hit it. So, I drew a straight line to the pond, took bearing and we started to cut up and down hill, across drainages. After a half days cutting, we came to the pond on bearing with considerable relief."

By nightfall, they had made it to Laguna de la India. "The men all wanted to return home," White wrote in his report. "After considerable inquiry it developed that the men had no confidence in the guide, in fact, he himself was not quite sure he knew the trail. With no delays or bad luck, according to his story, we could reach the seepage for which we started in three days. This meant three days to go and the additional time necessary to examine the seepage and return to the last settlement. Even on forced march it would have required a total of six days at a minimum. We had with the utmost care only provisioned for six days supposing that we were able to kill the same amount of game as had been our luck. Our ammunition was down to five rounds to the man and the trip led us into the Indian settlements. In the face of these arguments it seemed ridiculous to continue the expedition.

Florin W. Floyd and Horace L. Griley "at the drinking fountain in our thesis area," Breckenridge, Colorado, July 1915. E.W. Owen/Mirva C. Owen.

"Next morning we packed for the return," he wrote. "The oxen had played out so I gave my riding mule for packing and walked. It was hard going up one side of a ridge and down the other a lesser distance always climbing. The trail was ankle deep in mud. In mid-afternoon we reached a mountain village. I had never before known complete fatigue. When I relaxed my leg mussles (sic) I misjudged the position of the chair and landed in a heap on the floor.

"After a rest and a good meal I secured a fresh saddle mule and pushed on toward Valez hoping to reach a half way inn before dark. The mule train men remained to continue the next day securing what additional mules were needed. It rained en route, the trail became muddy and slippery and travel was reduced to a walk. I arrived at the Inn well after dark. I hammered on the door to awaken some one. Some one around the corner took a shot. I moved on. I later learned that they had trouble, or a bit of a fued (sic), with some neighbors and they thought they were being raided."

White took the main trail for Valez, which was wide and safe, constructed along the mountainside. "The night was pitch black, one could see nothing," he wrote. "My mule slipped, I reined her hard into the bank. Eventually we topped a hill and there were the lights of Valez. Then I knew. Instead of the wide trail I thought we were traveling, the mule had taken a short cut, a veritable goat path along the edge of a chasm that a couple of weeks earlier I had refused to ride in daylight."

Early in March, White visited the property of the Tropical Oil Company at Barrancabermeja and Cerro de Omir. "The Company have cleared a good trail to their camp and constructed a telephone line from there to Barrancabermeja," White wrote in his report. "I arrived at the Company's camp on the evening of March 8th and was very pleasantly received by their Manager, Mr. McCullough, who did all in his power to make me comfortable and gave me what information he could regarding their work....The camp consisted of warehouses, machine shop and a house for the manager and office force. The company also had a small steamboat, lighters and motor boats to transport supplies up the Colorado River to the drilling camps. Three wells had been started at one of the drilling camps, two of which had been abandoned. At the other camp, two wells had been started and one had already been abandoned. They had reached 600 feet when White was at the camp in early March, but by the end of the month, reports to New York indicated a very good showing of oil.

White moved on to Cerro de Omir, got little information and returned via river steamer to Barranquilla. He had had enough of South America for awhile. "The evening of the 16th by paying one of the passengers to stay behind I secured passage on the United Fruit steamer sailing for New York and arrived...March 29th."

White's official recommendation was disappointing, considering the time, the difficulty and the expense. His message to the company was one that was repeated by hundreds of geologists as the easy petroleum finds were discovered and the search grew more complex and more intense. White wrote, "No operations for the exploration of petroleum be undertaken or any of the properties in this district held under lease or promise of lease."

·ESTABLISHING A NEW SCIENCE·

In 1915, Dorsey Hager published Practical Oil Geology, one of the first books on the 'new' science. Hager wrote for the unschooled, but he taught the use of the transit, level and barometer and alidade. He discussed topographs, stratigraphs, folds, structures, exposures, surfaces, porosity, imperviousness, Permian and Cretaceous sediments.

The book even included a simple chloroform test for oil samplings of corings. His work was a major step in helping geology make the jump from a loose body of knowledge to an organized scientific profession.

The war contributed clothing and equipment to field geology, whether in the wilds of Oklahoma or the remote reaches of South America, as shown in this 1917 photo.   Mirva C. Owen.

Part of the geologist's job was to scout the country, but then he had to let others know about his discoveries. That usually included writing up reports. Many geologists, eager to impress their employers or clients, proceeded to pad their reports with geological terms and other impressive-sounding phrases. E.H. Cunningham-Craig was not impressed. "It should be the geologist's endeavour to try to see how short a report he can write, provided all essential matters are covered, and not how long he can make it," he wrote. "The geologist...should write out his report three times, each time making it shorter by cutting out all that does not seem absolutely necessary. Looked at from this point of view it is wonderful how much "padding" can be detected in even a workman-like and concise report. Perhaps one of the most fruitful sources of "padding" is in alluding to, discussing, or criticizing previous work done by others in the same area or district. This is very seldom necessary, except in the briefest possible fashion; it is wearisome to the reader, and it is occasionally dangerous....The last report *must* be the best, if the observer be competent, as he begins where his predecessors left off, with many of the essential facts already marshalled for him.

"Clearness is no less essential. Technical geological terms should be eschewed as far as possible, as it is probable that of those who read a report few will have more than a smattering of geological knowledge. It is not difficult to explain in simple language all that can be conveyed by sesquipedalian scientific phraseology. Again, it is not enough that the writer is clear in his own mind upon a point; he must set it down so that the reader cannot fail to be clear in *his* mind as to what is meant to be conveyed. This is not such a simple matter as it appears at first sight. In correspondence with reference to a report or the ground with which it deals, the geologist's statements will be paraphrased and unintentionally misquoted, and some day a statement which the writer considered impossible to misconstrue will come back to him distorted out of all recognition and labelled as his opinion. Therefore short, crisp sentences, without conditional clauses, should be the rule.

"Graces of style and the neat turning of phrases are to be avoided; it is possible to give a literary flavour to scientific work, as many of the greatest geologists, from Hugh Miller onwards, have taught us, but it is not literature that is required from the field geologist, but facts. If in reading over the draft of a report one comes upon any sentence with which one is particularly pleased, the wisest course is to cut it out at once. Be literal rather than literary.

"The point most essential of all is to stick to facts. Opinions *must not be given on any points of importance in the geology of the area examined.* It is, of course, impossible to avoid giving an opinion upon such a question as whether an area is sufficiently promising to warrant development work being undertaken or not, but in dealing with questions of structure, lateral variation, thickness of oil-bearing strata, depth to be drilled, etc., no mere opinion will suffice. If the certified facts cannot be given, the geologist must say so clearly. 'To the

A group interested in oil take a junket to Osage County prior to World War I. H.V. Foster (seated, far right) was head of ITIO and the initial development lessee of the entire Osage Nation. The woman standing fourth was Lois Straight Johnson, an early-day lawyer and lease document expert, married to Roy Johnson. Cities Service Company.

best of my belief,' 'as far as I could ascertain,' 'in my opinion,' 'it seems to me,' and the numberless similar phrases should be tabooed. Indeed, the geologist will do well to shun the use of the first personal pronoun as much as possible, and to write his report in the third person. The report will read better and will appeal more forcibly to both scientific and commercial readers if the writer does not intrude his personality, but allows the facts as ascertained by him and set forth in map, section and report to speak for themselves....

"Evidence of the presence of petroleum should be treated separately and at greater length, for much, and in some cases, perhaps undue, importance will be attached to such evidence by those for whom the report is written. It is always necessary to prove as conclusively as possible the petroliferous nature of the series that has been studied geologically, and the conditions under which surface shows of petroleum occur afford very valuable hints to the expert or technical adviser and the field manager....

"It will frequently happen that the geologist in the course of his field work will establish, or obtain evidence about, some point of general scientific interest, and he will naturally be tempted to enlarge upon it in his report. In such cases the best procedure is to consider whether the scientific point in question is of practical importance in the commercial development of any particular field, and whether other members of a scientific staff...will be helped in their investigations by the new knowledge acquired. If so, the evidence should be described briefly and the conclusion stated. Otherwise it is better not to overload a report with matters, however interesting and important from the scientific point of view, that have no direct bearing upon the practical finding and producing of petroleum."

· On the Athabasca ·

In 1913, Sidney Clarke Ells set out for Athabasca, Canada, to survey the prospects of commercial production for the federal mines branch. The area had been noted for over a century for its tar sands and pools. He was there to see if there were more than just surface traces. Conditions in the area were difficult. Ells had to travel by foot the 250 miles from Edmonton to Fort McMurray. He carried a 70-pound pack 250 miles over trackless muskeg and forest and camped out at night in temperatures as low as 50 below. He set out from Athabasca Landing north of Edmonton with a 30-foot scow, a 22-foot freight canoe and a crew of three white men and an "alleged" native pilot. It took nine days to float down the Athabasca the 240 miles to Fort McMurray.

In the following three months, they located 247 tar-sand outcrops extending over a distance of 185 miles along the banks of the Athabasca and its tributary rivers. They collected more than 200 samples from hand-augered holes that ranged from five to 17 feet. The return trip was weighted down with the samples, and there was nothing but manpower to handle them. For 23 days, Ells and a 12-man crew of natives pulled on a track line 20

Everett Carpenter was chief geologist in 1916. Seated, front left to right: W.E. Bernard, Walter Berger, Everett Carpenter, Roy S. Hazeltine, Fritz Aurin, Dean Stacy. Middle row: George Morgan, Vern Woolsey, Frank Parsons, W.C. Giffen, unknown. Back row: George Burress, Allen Bowden, L.C. Snider, Hun, George Mayor. Missing: J. Russell Crabtree. Cities Service Co.

hours a day, guiding the heavy-laden boats down the rivers. Only by sheer determination did they manage to bring out the first meaningful tar sand samples.

Ells was back in the country in early winter 1915. This time he had horses to help haul out the 60 tons of tar sands from McMurray to Edmonton. Conditions were still difficult. The temperatures ranged from 20 to 50 below zero, and there were no tents for men or horses. The tar sands were eventually used for experimental paving of Edmonton streets.

(Opposite and above)
Geologists gather at the International
Geological Congress in Toronto, Canada,
in 1913. Over 450 members were present.
American Heritage Center, University of
Wyoming.

Railroad cuts were a favorite place to examine geology, just as they are today (as are interstate roadcuts). The rocks were sliced and exposed, the breaks were fresh, and you didn't have to fight your way through tangle and brush to get there.

(Opposite)
Pence Rock, Canada, a sight shared by geologists at the International Petroleum Congress in 1913. American Heritage Center, University of Wyoming.

Arisay Beach, Nova Scotia, International Petroleum Congress in 1913. American Heritage Center, University of Wyoming.

**International Geological Congress
examines the peneplain from St. Anne
Mountain, Pence, Canada.   American
Heritage Center, University of Wyoming.**

**Women geology students pose in a camp
in the Arbuckle Mountains in 1910.
Oklahoma Historical Society.**

·Geologist As Lunatic·

Some areas of the United States seemed less promising than others. K.C. Heald once remarked, "A geologist advocating a search for oil in Idaho is apt to be classed either as a mild lunatic or as a joker with a distorted sense of humor."

·Gas No Matter What·

Alberta's first homesteader, John George (Kootenai) Brown, tapped an oil seepage on Cameron Creek in 1885. He dug pits and trenches, skimmed up the oil and sold it to nearby ranches as lubricating oil for a dollar a gallon. In 1899, he obtained the first government oil leases issued in Alberta.

The Canadian Pacific Railway watched as Brown and others like him kept coming up with oil. They were eager to find their own resources and set about a drilling campaign. In 1883, they drilled in southern Alberta for water for their locomotives and struck natural gas. In 1890, they drilled near Medicine Hat for coal and found natural gas. Subsequent water wells drilled at Brooks, Bassano, Dunmore and Bow Island continued with the same luck—no water, only gas.

·First "Lady" Geologist·

The University of Wisconsin was an early advocate of geological education and began graduating young men in the 1870s with mining and metallurgical degrees. The first graduate with a degree from the Department of Geology and Mineralogy was Charles R. Van Hise, who received his bachelor's degree in 1880.

In 1882, Miss Florence Bascom, daughter of John Bascom, the university president, graduated with two degrees—one in arts, the other in letters. The bright young lady came under the influence of Van Hise and grew interested in geology. In 1884, she received a third degree—in geology. Three years later, she received her Master of Science degree for a study of the sheet gabbros of the Lake Superior region. She went on to John Hopkins University where she became the first woman to receive a Ph.D. degree, the first woman to receive a Ph.D. in geology in America and the first woman Fellow of the Geological Society of America.

In 1895, she founded the department of geology at Bryn Mawr College and spent most of her life teaching there, interspersed with summer work for the U.S. Geological Survey. She was the first woman to be employed by the U.S.G.S. and continued working summers for them for nearly 30 years.

In February 1917, The Southwestern Association of Petroleum Geologists was established at Tulsa. The first officers were J. Elmer Thomas, president; Alexander Deussen, vice-president; Maurice G. Mehl, secretary-treasurer; and Charles H. Taylor, editor. The first official membership list, published in 1917, carried the names of 94 geologists. Most were working in the Mid-Continent and Gulf Coast districts. They were employees of about 12 oil companies or federal and state surveys, consulting geologists and university teachers. Others around the world liked the idea, and a year later, the group was renamed The American Association of Petroleum Geologists.

The real creators of the American Association of Petroleum Geologists, according to Wallace Pratt, were Everette L. DeGolyer, J. Elmer Thomas and Charles H. Taylor. "All the rest of us were there, of course, at one or more meetings," Pratt explained at the AAPG semi-centennial in 1966. "But these three men were the true Founders....I watched them in action at that time...DeGolyer, particularly, and Thomas; they were youth with a cause. But theirs was not the zeal of the Christian crusader; it was instead the zest of the perfectly oriented, supremely intelligent mind. They literally animated their project. Taylor, no less essential but older (a little) and more sober, guided, counseled, organized and gave form to the effort.

"There were, of course, other Founders whose contributions were indispensable: in Texas Bill Wrather and Alex Deussen; in Oklahoma Sidney Powers and Charles Decker; and in Kansas, Ray Moore. And there were still others—perhaps two dozen in all, among them fine scientists, A.W. McCoy, W.A.J.M. van der Gracht, Johan A. Udden and David White—whose assistance was of pivotal significance. But all these were supporters rather than founders. They supported the movement and accelerated its pace, once the true Founders had breathed into it the spirit of life."

Prior to World War I, Cities Service initiated the first wide-scale geological survey by an oil company. It started in eastern Kansas and progressed up to the northern Kansas border, back down through Oklahoma, Texas, then New Mexico and South America. It began in 1913 with Charles Gould and expanded in 1916 to the group shown here in Bartlesville.

·EPILOGUE·

Reflecting back on the early years, founder Wallace Pratt put these events in perspective. "Petroleum geology in the United States first attained the status of a profession when... the American Association of Petroleum Geologists was organized..." Another founder declared, "At the time of the organization of the Association, the usefulness of geology in prospecting for oil had been established."

"And the enlightenment which came with...more comprehensive knowledge of the crust of the earth is as valuable to science and to society at large," Pratt emphasized, "as are all the additional oil and gas fields the geologist has found."

On February 10, 1917, a group of men
organized under the name of the
Southwestern Association of Petroleum
Geologists. The next year they changed
the name to the American Association of
Petroleum Geologists. AAPG.

"...*we are indeed of the earth...*"

Harlow Shapley

·SELECT BIBLIOGRAPHY·

Amerada Hess, letters. Donated to AAPG.

American Association of Petroleum Geologists Bulletin. From inception to 1984.

Bartlett, Richard A. *Great Surveys of the American West.*
Norman, University of Oklahoma, 1962.

Burke, Marilyn. "Charles Hares, Western Oil Finder,"
Pacific States & Rocky Mountain Oil & Gas Reporter, December 1965, p. 19-20.

Clark , Dorothy Lang. "Who the Devil is Charlie Hares?"
Petroleum Today, Summer 1966, p. 20-22.

Crump, Irving. *Our Oil Hunters.* New York, Dodd, Mead and Co., 1948.

Cunningham-Craig, E.H. *Oil-Finding.* London, Edward Arnold, 1921.

Eby, J. Brian. *My Two Roads.* Houston, Gulf Publishing Company, 1975.

Facts and Fantasies of the Oil Patch. Desk and Derrick of Oklahoma City, 1975.

Fanning, Leonard M. *The Rise of American Oil.* New York, Harper & Bros, 1936.

Fohs, F. Julius and James H. Gardner. "Geology in Oil Finding,"
Fuel Oil Journal of Houston, Texas and Tulsa, Oklahoma, February 1914.

Gentz, Julie. "Scotland: Birthplace of Geology." *AAPG Explorer* (June 1984), p. 12.

Gould, Charles. *Covered Wagon Geologist.* Norman, OU Press, 1959.

Hamilton, Charles Walter. *Early Day Oil Tales of Mexico.*
Houston, Gulf Publishing Co., 1966.

The Humble Way. May-June 1957. Vol. 13, No. 1.

Knowles, Ruth Sheldon. *The Greatest Gamblers.*
Norman, University of Oklahoma Press, 1978.

Magnolia Oil News. Magnolia Oil Company. April 1931.

McLaurin, John J. *Sketches in Crude Oil.* Harrisburg, Pa., 1896.

Merritt, J.I. "Clarence King, Adventurous Geologist." *American West.* July/August 1982.

Owen, Ed. *Trek of the Oil Finders.*
Tulsa, Oklahoma, American Association of Petroleum Geologists, 1975. (Memoir 6)

Pacific Oil World. August 1983, Vol. 75, No. 8, p. 8-26.

Pratt, Wallace E. *Oil in the Earth.* Lawrence, University of Kansas Press, 1942.

Rabbitt, Mary C. *Minerals, Lands and Geology for the Common Defense and General Welfare.* Vol. 1: 1870. Washington, U.S. Geological Society, 1979 (#024-001-03151-5).

Reinholt, Oscar H. *Oildom: Its Treasures and Tragedies.*
Philadelphia, David McKay, 1924.

Schruben, Francis W. *Wea Creek to El Dorado: Oil in Kansas 1860-1920.*
Columbia, University of Missouri Press, 1972.

Scott, Otto J. *The Exception: The Story of Ashland Oil & Refining Co.*
New York, McGraw-Hill.

Tait, Samuel W. Jr. *Wildcatters: An Informal History of Oil-Hunting in America.*
Princeton U. Press, 1946.

Tinkle, Lon. *Mr. De: A Biography of Everette Lee DeGolyer.*
Boston, Little, Brown Co., 1970.

Tyson, Carl I., James H. Thomas, and Odie B. Faulk. *The McMan.*
Norman, University of Oklahoma Press, 1977.

University of Wyoming, Petroleum History Collection.
Extensive collection, including letters, diaries, notes and unpublished manuscripts.

Wallis, William E. *"Oil is Where You Find It."* Notes.

Weeks, Lewis G. *"....a lifelong love affair".* Connecticut, 1978.

Woolnough, W.G. *"The Parliament of the Commonwealth of Australia 1929-1930-1931. Report on Tour of Inspection of the Oil-Fields of the United States of America and Argentine, and on Oil Prospects in Australia."* n.d., location, etc.

Wright, William. *The Oil Regions of Pennsylvania.* New York, 1865.

Youngquist, Walter. *Over the Hill and Down the Creek.*
Caldwell, Idaho, Caxton Printers, 1966.

·ACKNOWLEDGMENTS·

The following people and institutions generously contributed to this book with their time and energy, and often with the loan of photos, documents, and other materials. Many items were sent from personal collections, while others came from museums, libraries, and industry archives.

We gratefully acknowledge the following contributors for their invaluable help in preserving this segment of petroleum geology's rich and colorful past.

American Association of
Petroleum Geologists (AAPG)
Tulsa, Oklahoma

American Museum
of Natural History
New York, New York

Aminoil
Houston, Texas

Atlantic Richfield Archives
Los Angeles, California

Burton E. Ashley
Washington, D.C.

Floyd Ayers
San Antonio, Texas

V.H. Barnett
U.S.G.S.

A. Bennison
Tulsa, Oklahoma

R.A. Bishop
Denver, Colorado

The Blakey Group
Tulsa, Oklahoma

H.H. Bradfield
Dallas, Texas

George Hansen Collection,
Brigham Young University
Provo, Utah

Laurence Brundall
Santa Barbara, California

W.D. Chawner
Del Mar, California

Joe M. Clark
Fayetteville, Arkansas

Willard J. Classen
Menlo Park, California

Virgil B. Cole (deceased)
Wichita, Kansas

Lynn R. Cook
El Paso, Texas

Robert H. Dott, Sr.
Tulsa, Oklahoma

Mrs. Rolf Engleman
(the late Rolf Engleman's photos)
Oklahoma City, Oklahoma

E.K. Erickson
Boulder, Colorado

Exxon Corporation
New York, New York

N.L. Falcon
Chiddingford, Surrey, England

John T. Galey
Somerset, Pennsylvania

Donald Gould
Los Angeles, California

E.G. Griffith
Denver, Colorado

Warren Griss
La Mesa, California

Michel T. Halbouty
Houston, Texas

M.R. Hall
U.S.G.S.

Jeffrey Heyer
U.S.G.S.

R.E. Horton
U.S.G.S.

William H. Jackson
U.S.G.S.

John S. Kelly
Midland, Texas

Robert E. King
Venice, Florida

H.M. Kirk
Carrollton, Texas

E.C. LaRue
U.S.G.S.

W.R. Moran
Los Angeles, California

National Archives
Washington, D.C.

I.K. Nichols
Houston, Texas

James A. Noel
Houston, Texas

Oklahoma Historical Society
Oklahoma City, Oklahoma

Western History Collections,
University of Oklahoma
Norman, Oklahoma

Harvey Olander
Thousand Oaks, California

Victor Oppenheim
Dallas, Texas

Mirva C. Owen
(the late Ed Owen's photos)
San Antonio, Texas

John M. Parker
Lakewood, Colorado

Samuel T. Pees
Meadville, Pennsylvania

L.J. Pepperberg
U.S.G.S.

Permian Basin Museum
Midland, Texas

Phillips Petroleum
Bartlesville, Oklahoma

Wallace Pratt (deceased)
AAPG Headquarters, Tulsa, Oklahoma

W. Armstrong Price
Corpus Christi, Texas

L.M. Prindle
U.S.G.S.

J. Herbert Sawyer
Los Altos, California

F.C. Schrader
U.S.G.S.

B.J. Smits
The Hague, Netherlands

R.W. Stone
U.S.G.S.

University of Missouri at Columbia
lent by Tom Freeman
Columbia, Missouri

Barker Texas History Center,
University of Texas at Austin
Austin, Texas

American Heritage Center,
University of Wyoming
Laramie, Wyoming

Cecil von Hagen
Houston, Texas

William E. Wallis
Santa Barbara, California

Roy C. Walther
New Orleans, Louisiana

Mrs. L.G. Weeks
(the late Lewis G. Week's Memoirs)
Westport, Connecticut

Sherman A. Wengerd
Albuquerque, New Mexico

Clayton W. Williams, Jr.
Tulsa, Oklahoma

Hugh Wilson
Houston, Texas

D.E. Winchester
U.S.G.S.

E.G. Woodruff
U.S.G.S.

A.F. Woodward
Tigard, Oregon

PRODUCTION ASSISTANCE: BOLEN, INK
LASER SCANNED DUOTONES: CHROMA-GRAPHICS, INC.
PRINTING: RODGERS LITHO
TYPE: COMPUGRAPHIC GOUDY OLD STYLE
PAPER: 80 # CONSOLIDATED CENTURA DULL
DESIGN: WINCHELL GRAPHIC DESIGN